GEOTHERMAL ENERGY UTILIZATION

EDWARD F. WAHL

Occidental Research Corporation
La Verne, California

A WILEY-INTERSCIENCE PUBLICATION
JOHN WILEY & SONS, New York · London · Sydney · Toronto

***Library of Congress Cataloging in Publication Data*:**

Wahl, Edward F.
Geothermal energy utilization.

"A Wiley-Interscience publication."
Includes bibliographies and index.
1. Geothermal engineering. I. Title.

TJ280.7.W33 621.4 77-546
ISBN 0-471-02304-3

Printed in the United States of America

10 9 8 7 6 5 4 3 2 1

Geothermal Energy Utilization

To Ed, Charlotte, my parents, and Ginny

Preface

Much has been written about the geochemistry, geophysics, geology, and general utilization of geothermal reservoirs. This book is directed at the process technology of geothermal fluids after they reach the surface. It presents information useful to engineers and scientists who are researching, designing, or evaluating process systems for making electricity, distributing thermal energy, or recovering minerals. Chemistry and thermodynamics are the foundations for describing geothermal resource utilization. A compilation of current knowledge about the process chemistry and thermodynamics of geothermal energy utilization is included together with a discussion of the basic theories and their interpretations. As such, the book will be useful both to the novice and the professional, as well as serving as a text for graduate or upper-level undergraduate courses.

CONTENT

Theories are presented in basic and complete form so that expertise is not required to use the information. The first chapter is an introductory description of geothermal energy covering those aspects important to utilization. The chemical and physical properties of brines are presented together with chemical and thermodynamic principles that relate to them. This information is necessary for engineering calculations as well as conceptual designs for geothermal brine process plants. The chemistry of deposition is described in detail and interpreted in terms of observed deposition in geothermal systems.

Thermodynamic principles are described in terms of geothermal energy utilization. These principles are then used to examine the performance of various expansion machines and process systems for producing electrical power. The thermodynamic limitations and inefficiencies become apparent from the analysis. These machines and systems are then compared with each other, with the ultimate and practical work output, and with space and process heat systems using the thermodynamic principles set forth earlier.

Data such as the physical and chemical properties of the brines and the thermodynamic performance of various power production systems are given both in graphic form and equation form. The graphic presentations are convenient for ease of calculation as well as for visualizing the relationships. The equations are useful for constructing mathematical models for computer computation. The equations also serve to identify the important parameters controlling a particular phenomenon.

The analysis of geothermal energy utilization systems is given in terms of the important parameters with all assumptions listed. Consequently, the reader is given an evaluation of the various prospects as well as example cases to assist in evaluating the readers' particular cases and any changes of assumptions. The information presented is useful both for gaining insight into geothermal energy utilization, and as a reference on chemistry and thermodynamics applied to this field.

Each chapter has its own reference list, which is further subdivided by subject matter where appropriate so that the reader may easily select additional readings according to his interest.

UNITS

The first four chapters deal mainly with geology and chemistry so that metric units are convenient and meaningful. On the other hand, Chapter 5 deals with application, so units more familiar to the practicing engineer are used. The last five chapters deal with thermodynamics, and so Btu, Btu per pound, °R, °F, and megawatts are used for the same reason. Each chapter has its own nomenclature list.

ACKNOWLEDGMENTS

The encouragement and assistance of the individual members of the Occidental Research Corporation staff in preparation of this book is deeply appreciated. In particular, I am indebted to Georgia Rowe, Jan Smith, Jannae Metz, and Jayne Flatt who typed the manuscript, particularly Georgia Rowe who did the tables; George Zyvoloski and Dick Price who took my related course at Cal Poly—Pomona and made many useful technical comments; Greg Lane who took care of the legal considerations with respect to the corporation; Shirley Jones for her encouragement; Ike Yen, my immediate supervisor; and Barrie Munroe who took his time to consider this matter for the corporation. The permission of the Occidental Research Corporation to publish this work is appreciated. The assistance of the corporation in providing the typing of the manuscript as well as the use of the library facilities is acknowledged.

Appreciation is expressed to Dr. Richard McKay of the Jet Propulsion Laboratory, Tom Fiorito and Scott Hector of Texaco, and Charles Priddy and Roger Martin of the California State Lands Commission for their useful comments on the text and associated course at California State Polytechnic University—Pomona.

Credit is due Geophysicist Dr. Alex Baird of Pomona College for reviewing the entire manuscript and particularly for his technical criticism of Chapter 1 and his assistance in rewriting it. Especially, I thank Fred Boucher for his thorough technical editing and many suggestions for improvement of the text.

EDWARD F. WAHL

La Verne, California
November 1976

Contents

Geothermal Energy Utilization

CHAPTER 1

Description of Geothermal Energy

CURRENT UTILIZATION

Because of the potential shown by past exploitation, along with predictions for the future, it is not surprising that governmental organizations throughout the world are making substantial research and development funds available for assisting and initiating geothermal energy utilization. Major corporations, as well as smaller ones, are actively exploring and purchasing leases for potential geothermal reservoirs. Others are pursuing the development of systems for converting geothermal resources into usable commodities. The result is that geothermal energy utilization will increase in the near future and probably for decades to come. Recent history of installed geothermal electrical production systems, Table 1.1, indicates that geothermal utilization is doubling about every 8 yr.

Geothermal energy is currently utilized in many parts of the world, not only for space and process heat for residential, commercial, and industrial uses as described in Chapter 9 but also for electricity generation. The 1975 power produced worldwide for each end use is 1300 MW, Table 1.2. Thermal energy for space heating and other purposes throughout a city such as that of Reykjavik, Iceland, costs

Table 1.1 Worldwide Geothermal Power Plants, Cumulative Installed Capacity in Megawatts

YEAR	ITALY	NEW ZEALAND	USA	MEXICO	JAPAN	ICELAND	USSR	OTHER	TOTAL
1913	1/4								1/4
1914	8								8
1936	56								56
1939	140								140
1940	250								250
1944	10								10
1956	260	69							329
1960	300		12						381
1961	320	181							501
1963	320		26						527
1965	340		54						575
1966					20				595
1967				3.5	33		5.7		617
1968			82						645
1969						3.4			648
1971	391		192						810
1972			302						919
1973			412	78.5					1104
1974									1104
1975		231	522		58				1289
1976			632			63.4		110	1569
1977			742	153.5	158			170	1915
1978			852						2025
1979			962						2135
1980		281	1072	183.5					2325
Source.	(1,12)	(2,10)	(3,11)	(4,5)	(6)	(7)	(8)	(9)	

(1) Villa, 1975
(2) Smith, 1970
(3) Lengquist, 1975
(4) Guiza, 1975
(5) B. Molina, 1970
(6) Axtell, 1975
(7) Ragnars, 1970
(8) Berman, 1975
(9) Berman, 1975
(10) Scholes, 1974
(11) McCabe, 1975
(12) Birsic, 1974

one-half that of a fossil fuel-fired heating system, a large portion of this cost being due to the distribution system. Thermal energy delivered in large volumes near a geothermal reservoir might cost as little as one-fifth that of fossil fuels. Electricity is currently produced from geothermal energy at a cost 25 to 30% less than that of coal, oil, or

Table 1.2 Geothermal Space and Process Heat Excluding Resort Bathing Usage[a] Compared with Electrical Power for 1975

Country	Electrical Power[b] (MW)	Thermal Power[c] (MW)	Source
Iceland	3	475	Einarsson, 1975
USA	522	30	Lund, 1974; Linton, 1974
Japan	58	37[d]	Komagata, 1970
USSR	6	300	Tikhonov, 1970; Dorov, 1974
Italy	391	1	Dragone, 1970
Hungary	—	363	Boldizsar, 1970; Belteky, 1972
New Zealand	231	70	Wilson, 1970; Burrows, 1974
Mexico	79	—	
Total	1290	1276	

Source:
[a] Bath-resort usage and the like may be 5 to 10 thousand thermal megawatts.
[b] From Table 1.1.
[c] Documented, actual use probably exceeds this by a considerable factor. A subsequently published compilation is in essential agreement with this thermal power data and explains the 5000 MW previously reported for the USSR (Peterson, 1976).
[d] In addition, hot springs discharging an estimated total of 1700 MW thermal power are used for recreational and health purposes.

nuclear plants. Geothermal energy deposits exploitable by current technology occur in most geologically active regions that are on land, Figure 1.1, for reasons that will become apparent from the discussions in this chapter. The exploitable geothermal energy reserves of the United States are estimated to be 1,000,000 MW-centuries of thermal energy. In addition to that, all hydrothermal reservoirs are estimated at 4,000,000 MW-centuries, and the total of all types of geothermal deposits at 10,000,000,000 MW-centuries (White and Williams, 1975).

To intelligently construct process plant–reservoir systems for utilizing such thermal reservoirs, knowledge of what they are, where they are located, how energy can be extracted from them, and how energy can be converted into the desired products is required. This chapter provides an overview of the nature and location of geothermal deposits as a basis for the discussions in later chapters on the extraction and conversion of geothermal energy as practiced presently and as is probable in the near future. In the long term, use of higher temperature geothermal deposits, such as active volcanoes, may require a substantially different approach than that of present practice.

SOURCE OF KNOWLEDGE OF THE EARTH'S INTERIOR

Knowledge of the interior of the Earth is based on deductions from indirect observations at its surface, with the exception of volcanic eruptions and drill holes. Volcanic eruptions carry material directly from within the earth to its surface, thus providing information on interior composition and, indirectly, on interior temperature and pressure conditions. From wells that have been drilled to depths of 5 km, a few to as much as 8 km, samples of the crust have been obtained and subjected to detailed chemical and physical analysis. A complex of geologic and geophysical observations of the earth suggest that the outer layers of the earth consist of seven or eight plates floating about on a thermally convective viscous mass. Separation and collision of such plates result in relative vertical motion of portions of the crust and mantle of the earth. Thus examination of the exposed surface of the earth reveals material that has come from interior layers in prior ages. Laboratory studies of such materials give information on the state of the material as it probably existed in the interior of the earth. Study of transmission of seismic waves due to explosions or natural earthquakes provides information on transition boundaries (and/or regions between solid–liquid phases), densities, and compositions of material within the earth. Indirect as these observations are, geophysicists have been able to establish a relatively detailed picture of the interior of the earth.

LOCATION AND DESCRIPTION OF GEOTHERMAL ENERGY

As understood today, the Earth consists of a thin (~30-km thick) alumino-silicate crust as a layer on a ferro-magnesium silicate mantle about 2800-km thick, and below which is a liquid iron–nickel core containing an inner solid core, Table 1.3. A boundary plate, the Mohorovicic discontinuity, exists between the 6- to 70-km thick surface crust and the mantle. The Mohorovicic discontinuity, Moho for short, is a plane that reflects seismic waves, indicating a density boundary. The lithosphere is the relatively rigid mass from the surface to a depth of 100 to 200 km. Below that is a viscous mass called the

Table 1.3 Temperature Distribution and Transition Zones Between Regions of Earth's Interior

Region	State	Connecting Zone	Depth (km)	Temperature (°C)	Density (g/cm^3)	Composition	Region
Crust			0	0–50			Lithosphere
	Rigid Plates		10–20[c]	?	2.7	Na,K alumino-silicates[a]	
					3.0	Fe,Ca,Mg alumino-silicates[b]	
		Moho	6–70	500–1000			
Mantle	Solid						
		Solidus	100–200	1200			
	Viscous Mass				3.6–4.4	Fe,Mg silicates	Asthenosphere
		Solidus	700	1900			
	Rigid Mantle				4.5–5.5	Fe,Mg silicates &/or oxides	Mantle
		Solidus	2800	3700			
Core	Liquid				10–12		Core
		Solidus	5500	4300		Fe,Ni	
	Solid				12–13		
		Center	6340	4500		Fe,Ni	

[a]Sial: Silicic crust - consisting of composition shown, non-existant in oceanic crust.

[b]Sima: Mafic crust - consisting of composition shown.

[c]Conrad discontinuity, 0 under oceans.

asthenosphere, which is continually circulating due to the thermal gradient between the interior and exterior of the earth. Note in the previous discussion and Table 1.3 that there is a dual terminology for the earth's zones. The one consisting of crust, mantle, and core is the original designation; the other is more recent and results from the plate-nature of the outer mantle and crust.

The oceanic crust is 6- to 10-km thick and consists of basaltic material. This crustal layer extends beneath the continental crusts as shown in Figures 1.2 and 1.5.

The continental crust, which consists of a low density granitic material that is predominantly sodium potassium alumino-silicates, is typically 35-km thick but may reach depths of 70 km in regions of mountain building. The greater thickness of the continental crust means that it is a better insulator for magmatic, or thermal, deposits than the oceanic crust. The conduction of thermal energy through the crust of the earth is such that a locally deposited magmatic intrusion will take several million to tens of millions of years to cool. Thus the life of a continental thermal deposit is about ten times that of the largely volcanic oceanic crust. And, as shown in Table 1.4, the temperature at a depth of 40 km in geologically active areas is 1000°C compared with 500°C in geologically inactive areas, such as the east-central United States. An average temperature distribution and density through the regions discussed previously are given in Table 1.3. Because of the circulation within the earth, as well as material

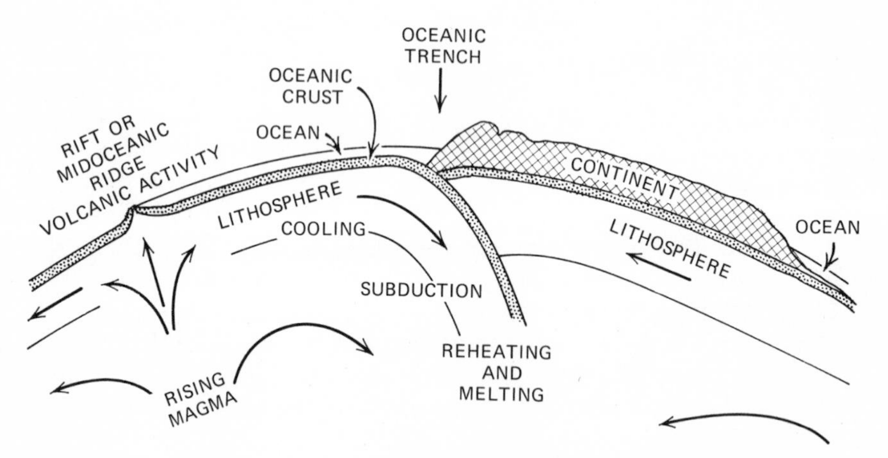

Figure 1.2 A model of the crust and mantle of the earth showing movements of the plates and lithosphere and thermal convection of the viscous mass beneath.

Table 1.4 Geothermal Heat Characteristics for Various Geologic Crustal Types

		Surface Flux	
Geologic Crustal Type	Temperature 40 km deep (°C)	Total (μcal/cm^2 s)	Radioactive (% of total)
Geologically inactive, e.g. East-central U.S.	500	1	30
Geologically active, e.g. mountain building, volcanism	1000	2	30
Oceanic crust at ridge of separate plates	--	>2	<5
Oceanic crustal plate	--	1.3	<5
Oceanic crust at trenches	--	<1	--

differences, the temperature of, and interfaces between, the regions vary spatially from the typical values given. The temperature distribution through the earth is due to the flow of heat from the stored thermal energy in the interior to the cold surface, plus any locally produced thermal energy. The stored thermal energy at high temperature within the interior of the earth produces a local flux at the surface of 1 to 2 μcal/cm^{2}s due to conduction from the high temperature interior through the crust to the colder surface.

One source of locally produced thermal energy that has been identified is that produced from radioactive elements, especially in granitic material. A 10-km thick granitic crust will generate enough radioactively produced heat to cause a thermal flux at the surface of 0.6 μcal/cm^{2}s. This radioactively produced heat is significant relative to the energy stored within the interior. The radioactively produced heat in basalt and peridotite, which are the more common components of the oceanic crust, are much smaller as shown in Table 1.5. The heat generated radioactively in the surface crust, though significant, is distributed over a relatively large volume and so is available at a relatively low temperature. Therefore, it is less useful than thermal energy in the interior of the earth that is at a higher temperature.

The work resulting from plate collisions may produce locally high

Table 1.5 Radioactively Produced Energy in Various Crustal Materials

Material	Energy Per Unit Mass (μcal/g yr)	Energy Flux from 10 km Thick Crust (μcal/cm^2 s)
Granite	7	0.6
Basalt	1.2	0.1
Periodite	0.02	0.004

temperatures sufficient to cause partial melting. This phenomena, anatexis or ultra-high temperature metamorphism, is another source of locally produced heat near the earth's surface.

The thermal energy flux of 1 to 2 μcal/cm^{2}s at the surface of the earth due to conduction from the interior from all geothermal sources is about one-five thousandth of the solar energy flux, which is a marginal energy source. Thus use of this conductive flux as an energy source is not particularly practical. The energy source of most interest is the high temperature interior. One way to tap this energy would be to drill a well 50-km deep where temperatures are 500 to 1000°C. Unfortunately, techniques are not available for drilling to such depths economically. Another approach is to transport the energy from depths nearer to the surface, where it can be tapped more readily. Fortunately, naturally occurring circulatory currents in the mantle of the earth, together with the hydrothermal circulations within the crust, are such a transport mechanism. To locate and understand the significance of near-surface deposits so produced, it is important to know the mechanism of transport of geothermal energy from the interior to the surface.

TRANSPORT OF INTERIOR THERMAL ENERGY INTO THE CRUST

A simplistic model of the crust and mantle of the earth, shown in Figure 1.2, consists of a relatively low density continental mass as a "raft" on a rigid lithosphere, which in turn "floats" on the asthenosphere. These continental crustal rafts move in various directions due to the convective currents in the asthenosphere and collide with or separate from oceanic plates or other continental rafts. These regions of interaction are the geologically active areas in which volcano eruption, mountain building, one plate subducting under another, and/or one plate folding on top of another occur. This activity causes hot material, that is magma, to intrude into the surface crust. These intrusions are caused by three types of plate interactions.

Plate Separation

Separating crustal plates being moved apart by convective currents in the asthenosphere result in magma flowing upward to produce volcanic action and provide hot material to the lithosphere as shown in Figure 1.2. As this plate is generated and moves along the surface, it cools and eventually collides with a continental plate or another oceanic plate. Due to this movement of the oceanic crust, the local heat flux varies with the location of the crust. It is about 1.3 μcal/cm^{2}s over its general surface area but less than 1 μcal/cm^{2}s in areas of trenches where the crust is subducting beneath a colliding plate. Due to the flow of magmatic material directly up through volcanic areas, the local heat flux near ridges is larger than from the general plate area, fluxes as high as 7 μcal/cm^{2}s having been measured, Figure 1.3.

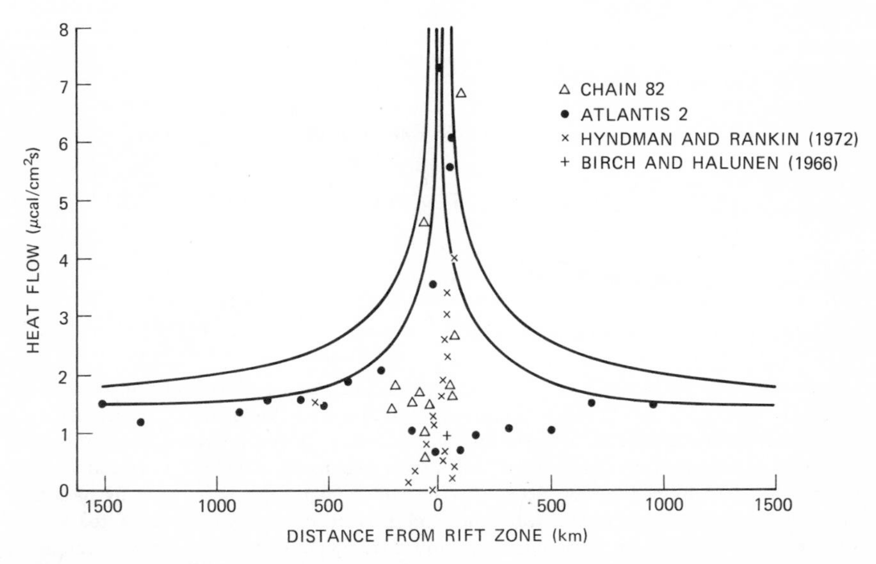

Figure 1.3 Heat flow profile across the mid-Atlantic ridge, 43 to 46°N. Solid curves are the theoretical profiles of Kasameyer et al. (1972), for constant spreading rates of 0.75 and 2.0 cm/yr. From Foster et al. (1974).

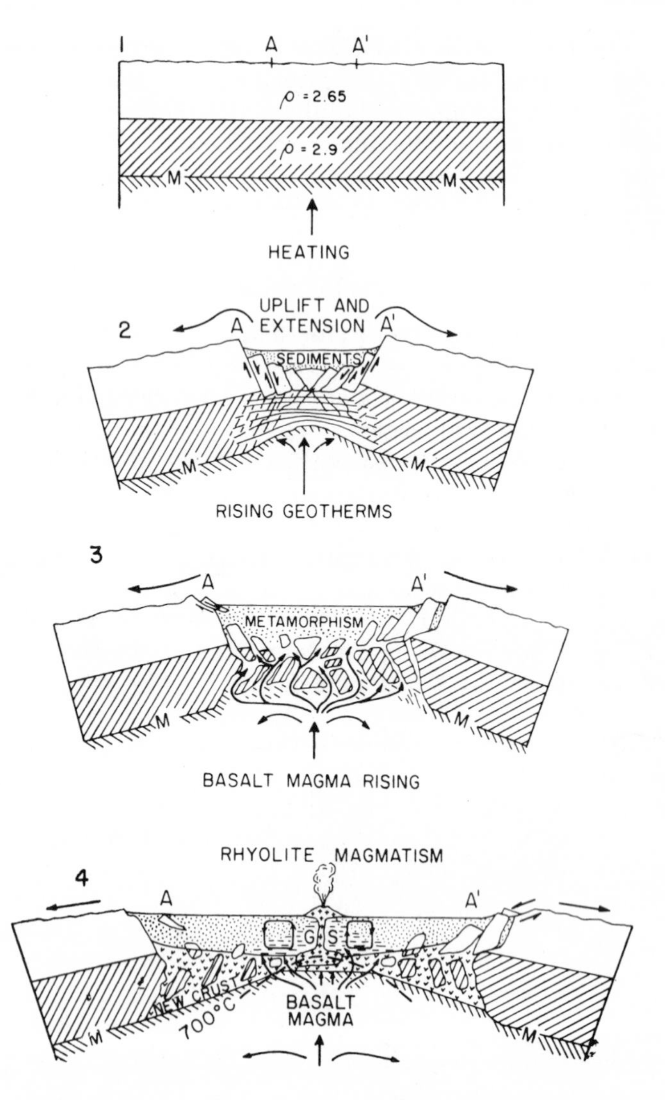

Figure 1.4 Model of rifting and magma generation during growth of a graben. The sections are drawn parallel to strike–slip faults. Stage 1—Two layers of crust overlie a hot zone in the mantle. M, Mohorovicic discontinuity; A and A′, reference points for later movements. Stage 2—Upward and lateral expansion; a trough is initiated and partly filled in by sediments. Stage 3—The widening trough is invaded by basaltic magma; metamorphism of the sediments and gravitational sliding of the tilted walls occur. Stage 4—Melting of the basement and extrusion of rhyolitic magma. Ascending hot brines cause greenschist metamorphism (GS) at shallow depth. From Elders et al., *Science*, Vol. 178, pp. 15–24, Figs. 5 and 6, 1972. Copyright 1972 by the American Association for the Advancement of Science.

The lifetime of an oceanic crust from its generation to subduction is on the order of one hundred million years. An example of crustal spreading and a region of active geothermal exploration and exploitation is the Imperial Valley region of southern California. In this region, the crust has thinned to 19 to 24 km (Koenig, 1967). A model for the formation of this structure is shown in Figure 1.4.

Plate Collision—Subduction

Oceanic plates colliding with continental plates will flow downward under the continental plate as shown in Figure 1.5. This subduction typically results in oceanic trenches. Such a geologically active zone results in magmatic intrusions due to partial melting and upward transport of the lithospheric material. These intrusions usually manifest themselves as volcanoes.

Plate Collision—Obduction

Another type of plate collision is characterized by folding and thickening of the crustal mass as one plate moves up over another plate. This geologic activity, called obduction, produces magmatic deposits as a result of metamorphism and/or intrusion of batholithic masses. The Himalayan mountains and Sierra Nevada are examples of the result of such activity.

The three processes already described occur in regions where the convective flows in the athenosphere are changing direction relatively radically. Consequently, small eddies would be expected that would contribute to magmatic intrusions in these regions. These geologically active regions are also sites of valuable mineral deposits because of the activity that creates them as shown, for example, in Figure 1.6. This may be of consequence because close proximity of magmatic deposits to mineral deposits that require process energy for extraction or conversion may make the two, when combined, more economically attractive for exploitation.

As described, the convective currents in the earth's mantle transport the thermal energy from the earth's interior to regions nearer the surface of the crust where they are deposited and insulated with a lifetime of several million years. These deposits in geologically active areas represent the most easily tapped heat sources because of their closeness to the earth's surface. Even so, their depth is sufficiently great that they are usually beyond the depth of economic drilling. However, the surface of the earth's crust contains free water and this

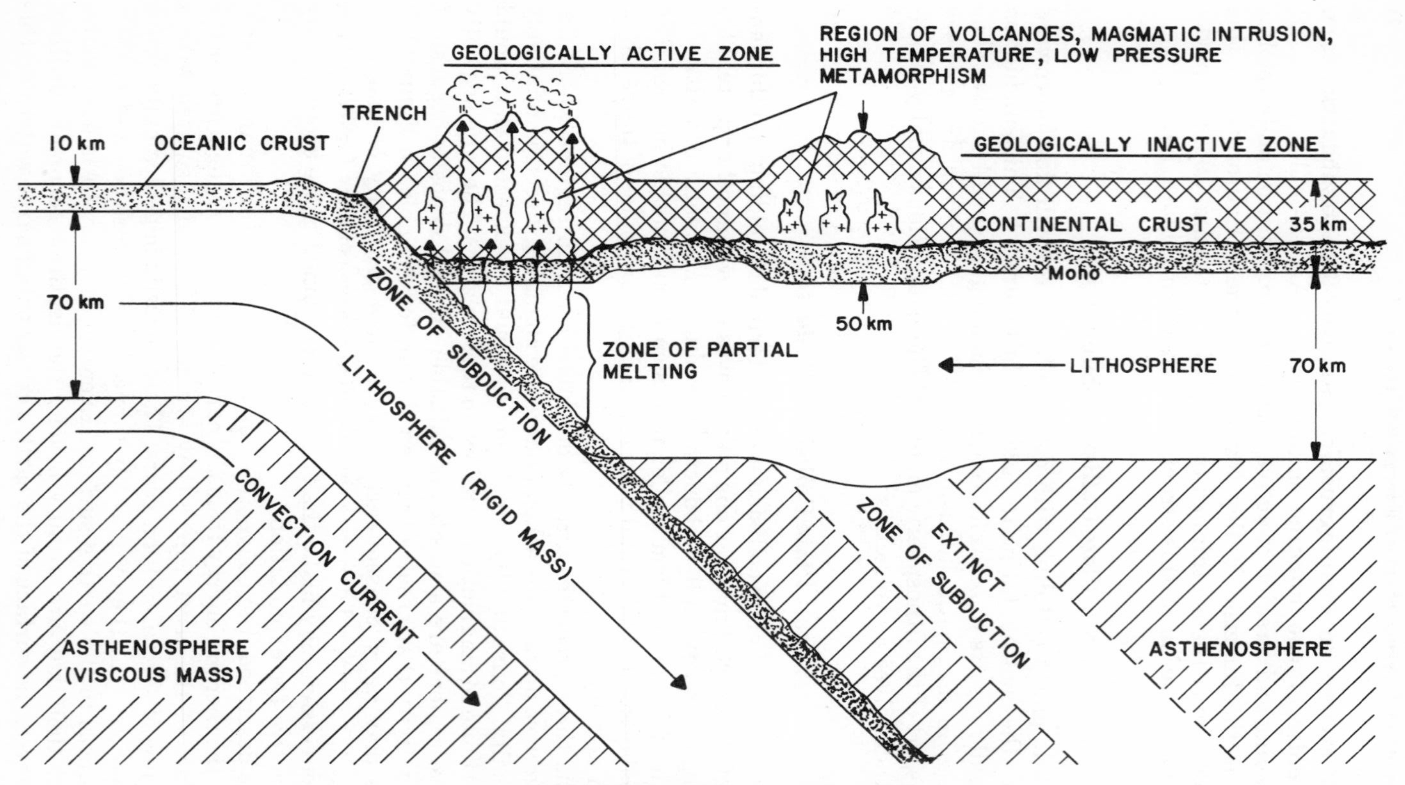

Figure 1.5 Plutons formed by subduction.

water sometimes circulates near magnetic deposits and transports some of the thermal energy near or through the surface of the earth. This is manifested on the surface as hot springs or geysers.

HYDROTHERMAL TRANSPORT OF GEOTHERMAL ENERGY

The outer crust of the earth is made up largely of sedimentary material, often layered in strata. These strata will have different permeabilities to water owing to different compositions, textures, and history. Some are highly permeable as a result of having been formed from coarse material or having been fractured. There are also strata that are impermeable to water. The result is that the meteoric water* in the upper layers of the crust flow in various patterns through more permeable zones. In some cases, oceanic water will permeate the crust. A very small fraction of the water, less than 5%, originates from the solid material itself, that is, from the magma that has come up from the interior portions of the earth. In regions where there is a pluton, that is, a magmatic intrusion into the crust of the earth, the water near it will be heated, and as a result, convective water currents will be set up that take the heat from the plutons and carry it upward and outward to distribute it throughout the crust. These convective currents vary in size from small isolated regions where the circulation is limited by the rock structure to regions up to 30 km in diameter.

By using the ratio of deuterium to hydrogen and the oxygen-18 to oxygen-16 ratios, geologists have been able to deduce extinct hydrothermal currents in the regions of magmatic intrusions that have already cooled, thus identifying "fossil" hydrothermal convective currents. Some examples from the work of Taylor are shown in Figure 1.7.

The size and shape of the convective currents as well as their pressure, temperature, phase distribution, and salt content will be as varied as the geologic structures in the crust of the earth, because it is the geologic structures that determine these characteristics. Nevertheless, it is useful to have some models to represent typical convective flows. Two such simplified models of convective currents that carry the heat from a magmatic source to the surface of the earth where it can be tapped and used are shown in Figure 1.8. Hot water reservoirs such as these occur, for example, at the Wairakei field in

*Water of atmospheric origin.

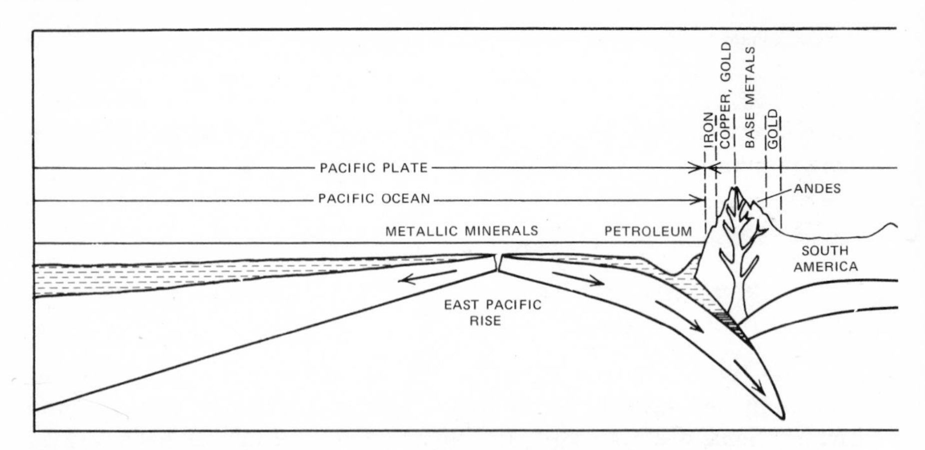

Figure 1.6 Mineral deposits in geologically active regions. Organic matter, salt, and metallic minerals accumulate in regions of spreading oceanic crust. The trapped organic matter forms oil and gas. Melted metallic minerals intrude into the crustal regions due to converging crustal plates (Press and Siever, 1974). Hydrothermal

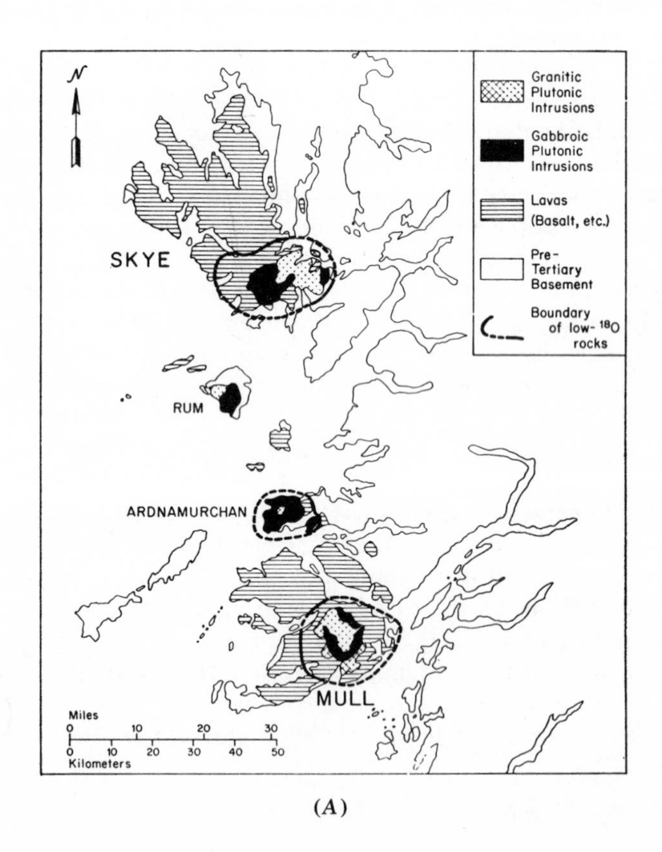

(*A*)

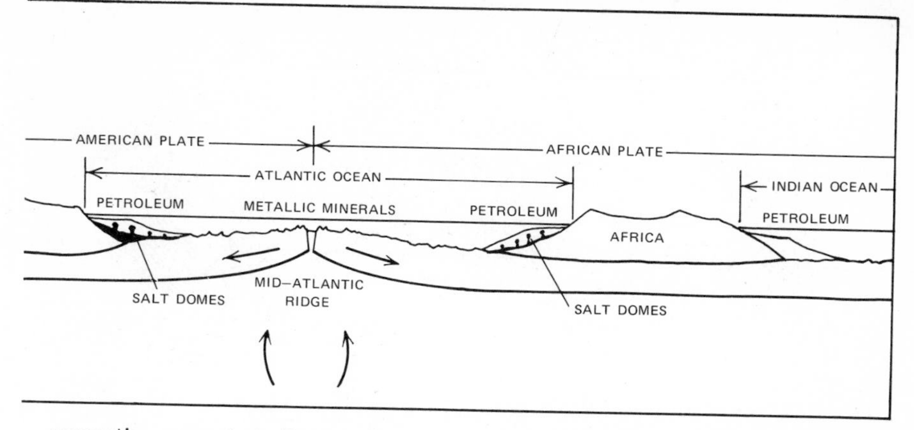

convection currents in these regions also concentrate metallic sulfides in convergent plate regions and concentrate first metallic sulfides then later metals in divergent plate regions; see Figure 1.15. From "Plate Tectonics and Mineral Resources" by Peter A. Rona. Copyright © July 1973 by Scientific American, Inc. All rights reserved.

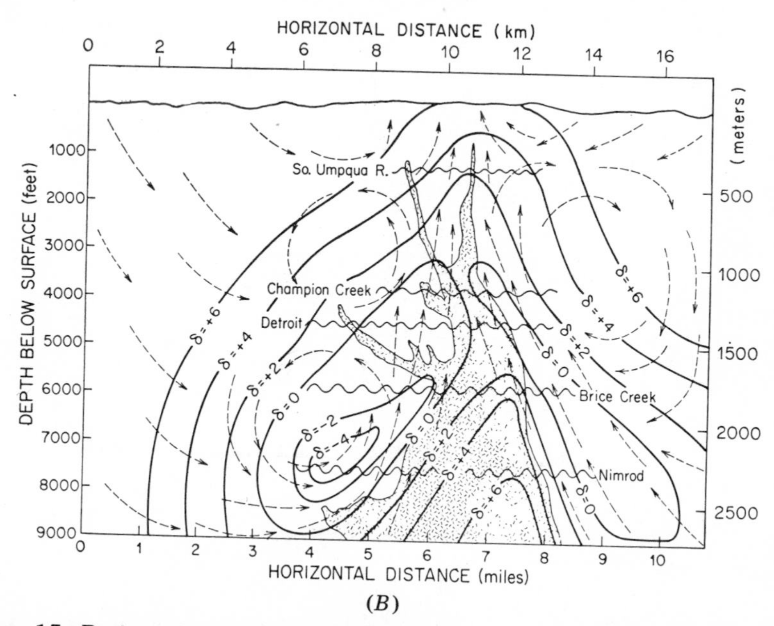

(*B*)

Figure 1.7 Regions of interaction of meteoric water with igneous intrusions showing the extent of the convective hydrothermal circulations. (*A*) Inner Hebrides of western Scotland showing the extent of oxygen-18 depletion. From Taylor (1974): *Economic Geology*, Vol. 69. (*B*) Hypothetical granodiorite stock of the western Cascade Range showing the depletion δ of oxygen-18 and the deduced circulation of the water indicated by the arrows due to heating from the intrusive stock. From H. P. Taylor, *Journal of Geophysical Research*, Vol. 76, p. 7868, 1971. Copyrighted by American Geophysical Union.

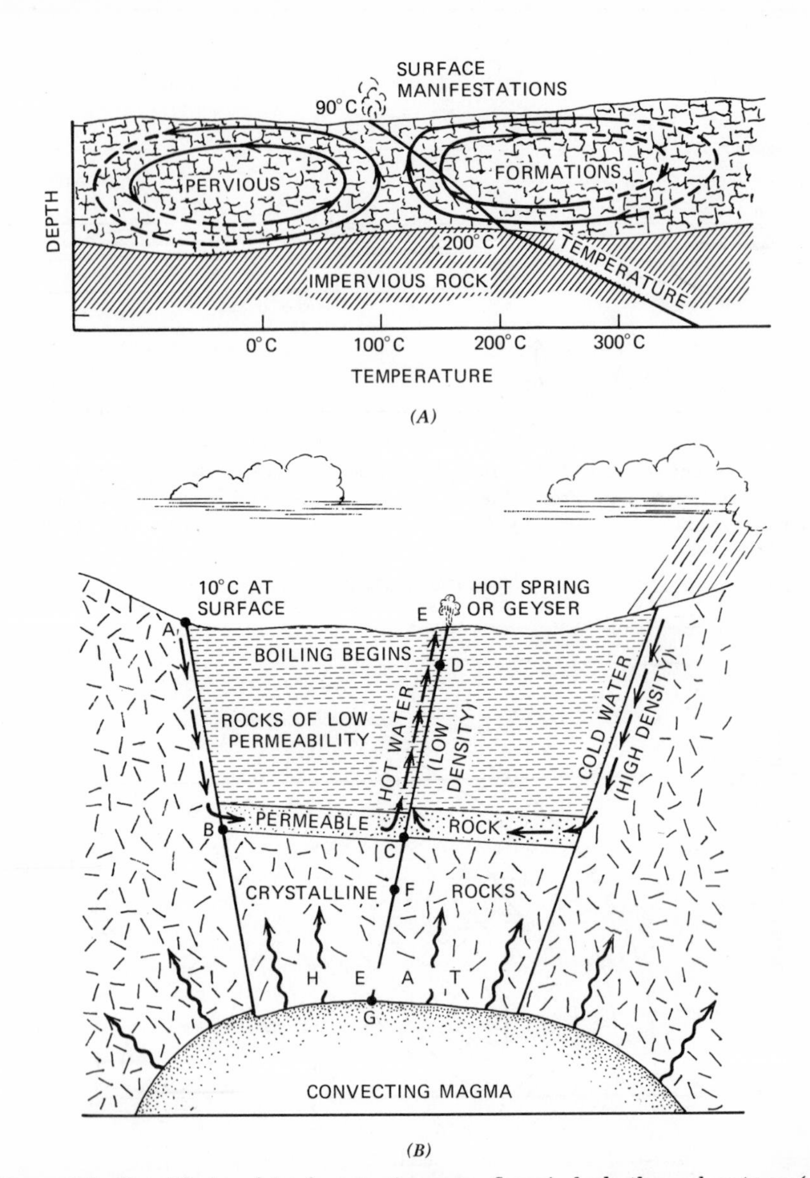

Figure 1.8 Simplified models of convective water flows in hydrothermal systems. (*A*) Basic flow showing temperature distribution in the rising water flow. From Facca (1973). Reprinted by permission of UNESCO from *Geothermal Energy*, Review of Research and Development, © UNESCO, 1973. (*B*) One effect of variations in permeability of the region on flow. Reprinted with permission from "Characteristics of Geothermal Resources," by Donald E. White, in *Geothermal Energy: Resources, Production, Stimulation*, edited by Paul Kruger and Carel Otte (Stanford: Stanford Univ. Press, 1973), Fig. 4, p. 76.

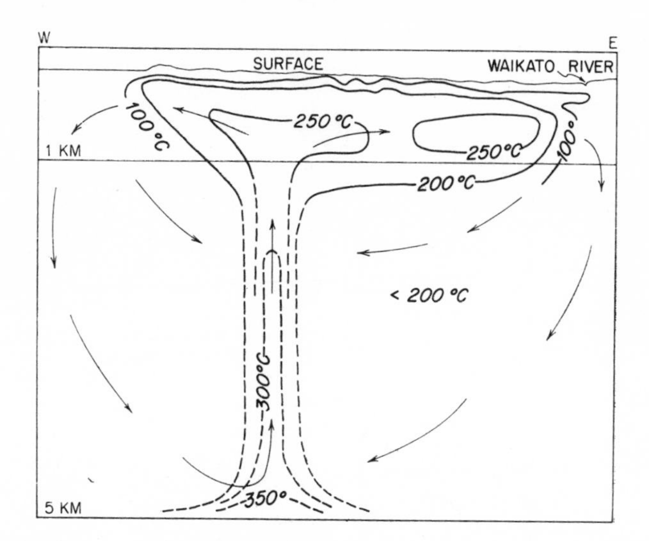

Figure 1.9 Vertical section through the geothermal reservoir at Wairakei, New Zealand, showing actual measured (solid lines) and estimated (dashed lines) isotherms to a depth 5 km beneath the surface. The approximate flow lines of the meteoric water are shown with arrows. From Taylor (1974): *Economic Geology*, Vol. 69.

New Zealand, which has the flow pattern shown in Figure 1.9., and Cerro Prieto, which has the flow pattern shown in Figure 1.10. In both of these fields it is possible to drill a well down through a hot zone and into a cooler zone. Since convective currents such as these carry the thermal energy to the surface of the earth more rapidly than it is carried by conduction, the temperature versus depth curve for such a system will be quite different from that of a pure conductive system. An ideal temperature plot is shown in Figure 1.11 for a convective cell such as the generalized flow model of Figure 1.8 compared with a purely conductive heat flow zone. Thus the presence of geothermal reservoirs at depth are detected or indicated by temperature logs that show greater than normal heat rise with depth indicating a convective zone carrying heat from depth. This type of circulating hot water reservoir is the one most frequently found. These hot water systems can be driven by conduction from a magmatic intrusion or by transport from a lower magmatic source by a feeder as shown in Figure 1.12. A postulated mechanism of a feeder for the Hakone geothermal system is shown in Figure 1.13. These flow models demonstrate the complexity and diversity of typical hydrothermal flows that are driven by temperature differences and that transport crustal geothermal energy upward toward the cooler surface of the earth.

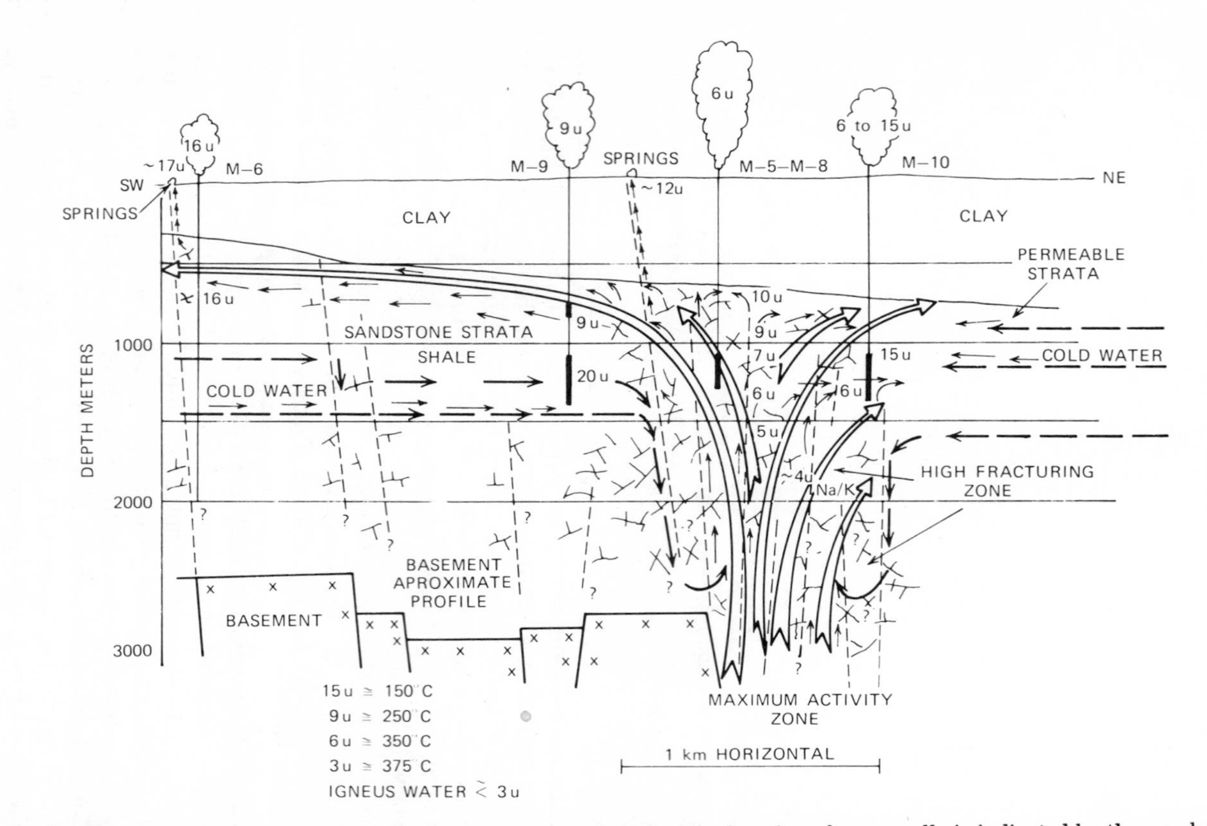

Figure 1.10 Water circulation in the Cerro Prieto geothermal reservoir. The location of some wells is indicated by the numbers M-5, M-8, and so on. The temperatures are deduced from the Na/K ratios as shown. From Mercado (1970).

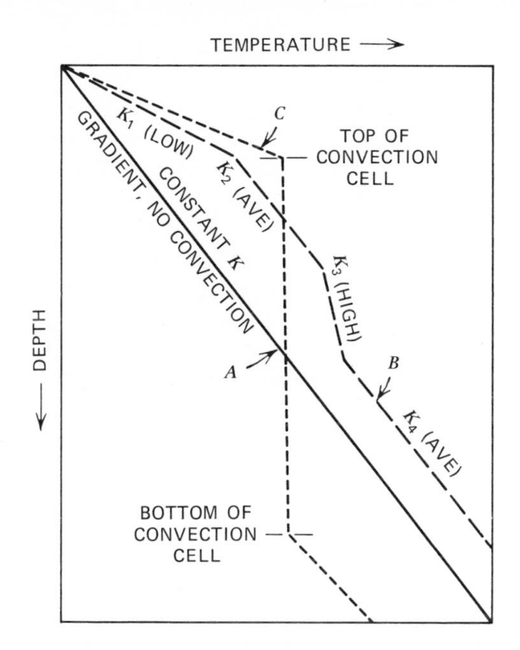

Figure 1.11 Temperature/depth relations where heat flow is controlled by thermal conduction in rocks of constant conductivity (A) or rocks of variable conductivity (B) or by major convective disturbance (C). Reprinted with permission from "Characteristics of Geothermal Resources," by Donald E. White, in *Geothermal Energy: Resources, Production, Stimulation*, edited by Paul Kruger and Carel Otte (Stanford: Stanford Univ. Press, 1973), Fig. 2, p. 72.

Another type of hydrothermal system is the so called dry steam field. A model for such a system is shown in Figure 1.14. In this system, the water is heated to a temperature that exceeds its boiling point at the static pressure of the system so that the water boils to form steam. Note that it is theoretically possible to draw the water out of a water system at such a high rate that the reservoir pressure is lowered to a point where the water flashes in the reservoir forming steam. In the actual operation of reservoirs, flashing will occur to varying extents, depending on the production rates.

A third type of hot water resource sometimes encountered is the so-called geopressured system. In this system, the heated water is trapped in layers in which the pressure exceeds that of the hydrostatic pressure due to compression of the zones above the layer. Such systems are found in the gulf states and in the Gulf of Mexico where gas and oil are also present in the reservoir along with the heated water.

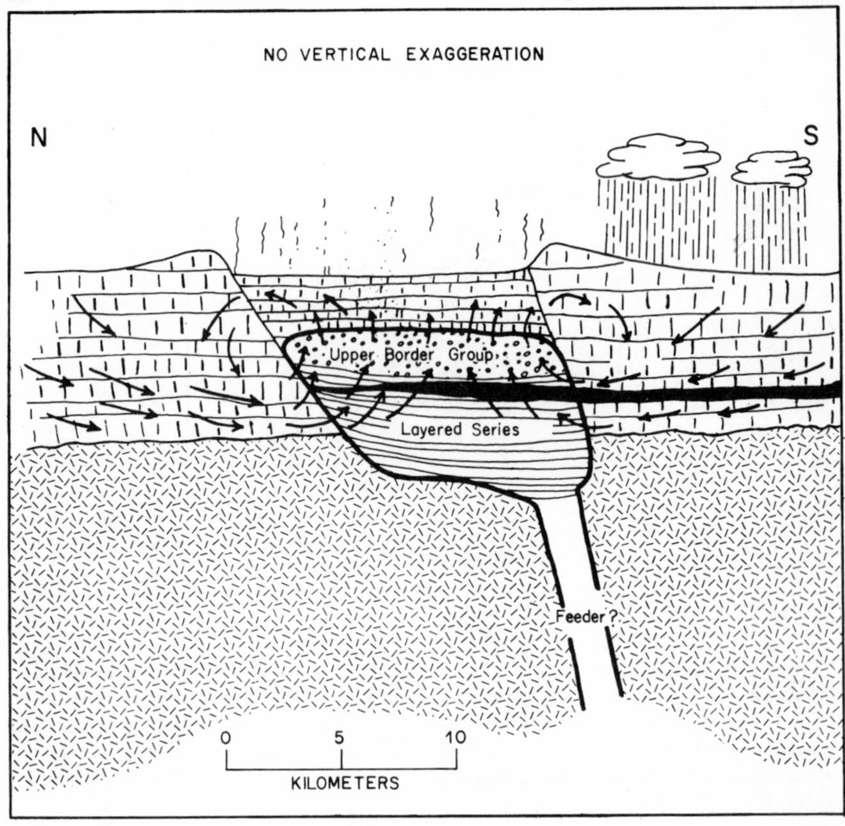

Figure 1.12 Hydrothermal circualation of the Skaergaard intrusion in the plateau lavas above the granite–gneiss basement. Transport of thermal energy through a feeder from a lower magmatic source may have been the energy source. Little meteoric water penetrated the basement. From Taylor (1974): *Economic Geology*, Vol. 69.

In general, hydrothermal waters will be 95% or more of meteoric origin. In some cases, there may be a larger amount of oceanic water. An example of the flow of oceanic water into the heated zones is demonstrated by a correlation of the oceanic tides with the pressure or water level in some of Iceland's geothermal wells. But in all cases, the amount of water from magmatic sources are at most only a few percent.

In all these systems, hydrothermal convective currents dissolve certain minerals in the hot zone and deposit them in the cooler zones. The result is the formation of mineral deposits such as copper, silver, and zinc. These mineral deposits are associated with older geothermal deposits as illustrated in Figure 1.15.

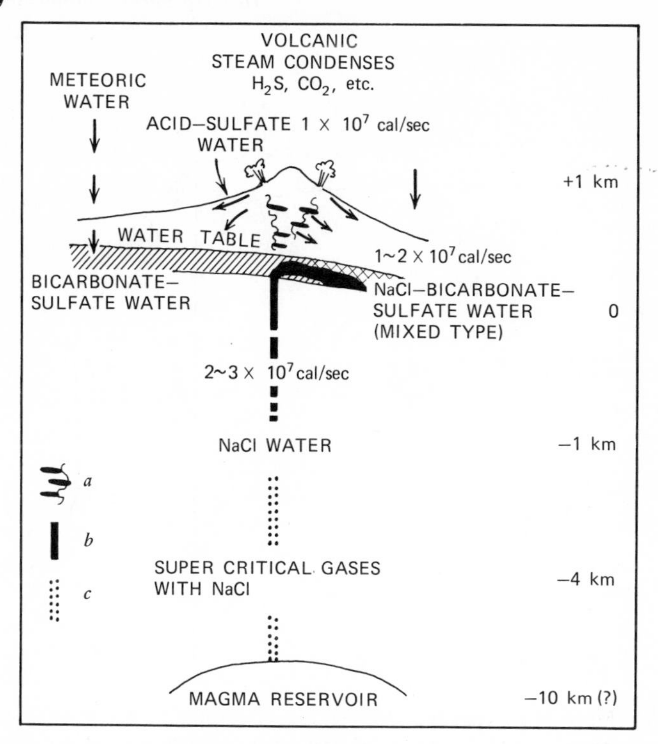

Figure 1.13 Possible mechanism for the genesis of the Hakone geothermal system. *a*, Repeated processes of vaporization and condensation of volcanic steam; *b*, hot sodium chloride water; *c*, super-critical gases with sodium chloride. From Oki and Hirano (1970).

The last type of geothermal deposit as identified for geothermal resource purposes is the so called dry rock system. This system corresponds to a magmatic intrusion, which is surrounded by, or is itself of, impermeable rock so that water cannot flow near and cool it. As a result, such a system retains its heat over a very long period compared to a hydrothermal system. It has been proposed that such structures can be hydraulically fractured to allow water circulation for recovery of the thermal energy.

SUMMARY

All indications are that exploration, research, development, and utilization of geothermal energy will increase considerably in the

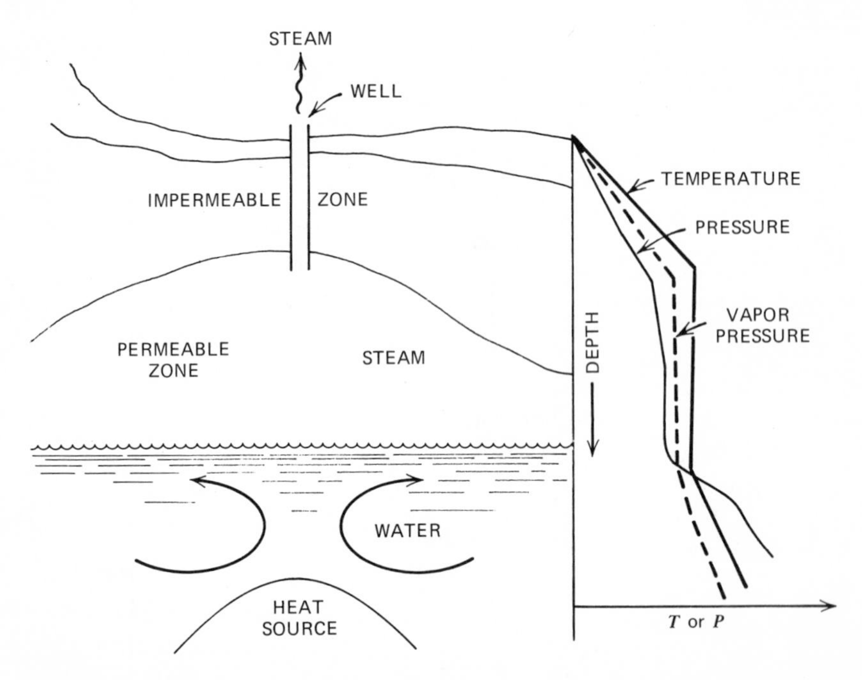

Figure 1.14 Model of a steam field showing the temperature distribution in the central portion of the system. In the upper region, the vapor pressure of water exceeds the local pressure so that the water is present as steam. In the lower region, the static pressure exceeds the vapor pressure so that the water is present in the liquid state.

forthcoming decade as evidenced by the increase in activity of federal agencies and private corporations, as well as the history of installed geothermal utilization plants. The source of geothermal energy is the earth's hot interior from which heat convects to the lithosphere about 100 to 200 km below the surface of the earth. This thermal energy just below the lithosphere of the earth is at a temperature of about 1200°C and is transported into the upper crust of the earth by convective eddies and magmatic intrusions in geologically active regions of the world where the crustal plates are moving together or spreading apart. These magmatic intrusions in turn are cooled by convective meteoric water in the permeable layers of the upper crust of the earth. Convective cells as large as 10 mi in diameter have been identified. The convective cells transport the thermal energy toward surface regions that can be reached by conventional drilling technology. These hot water deposits are the geothermal reservoirs presently being exploited.

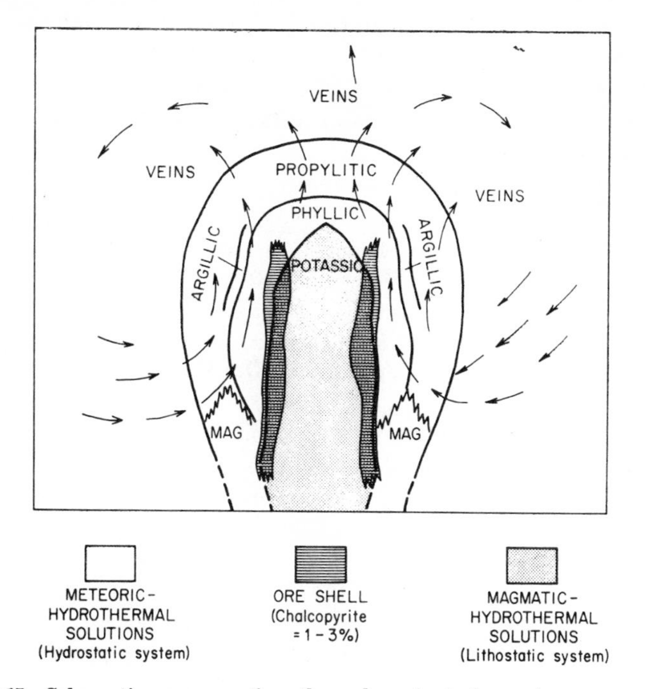

Figure 1.15 Schematic cross section through a typical porphyry copper deposit showing the two types of hydrothermal systems on either side of the deposit. From Taylor (1974): *Economic Geology*, Vol. 69.

REFERENCES

Elders, W. A., et al., "Crustal Spreading in Southern California," *Science*, **178**, 15–24 (1972).

Facca, G., *Geothermal Energy—Review of Research & Development*, UNESCO, Paris, 1973.

Foster, S. E., et al., "Heat-Flow Near a North Atlantic Fracture Zone," *Geothermics*, **3**, 3–16 (1974).

Haenel, R., "Heat Flow Measurements in Northern Italy and Heat Flow Maps of Europe," *J. Geophys. Res.*, **40**, 367–380 (1974).

Kasameyer, P. W., et al., "Heat Flow, Bathymetry, and the Mid-Atlantic Ridge at 43°N," *J. Geophys. Res.*, **77**, 2535 (1972).

Koenig, J. B., "The Salton-Mexicali Geothermal Province," *Calif. Div. Mines Geol. Mineral Inf. Serv.*, **20** (7), 75–81 (1967).

Mercado, S., "High Activity Hydrothermal Zones Detected by Nalk, Cerro Prieto, Mexico," *Geothermics*, Special Issue 2, **2** (2), 1367–1376 (1970).

Oki, Y., and T. Hirano, "The Geothermal System of the Hakone Volcano," *Geothermics*, Special Issue 2, **2** (2), 1157–1166 (1970).

Press, Frank, and Raymond Siever, *Earth*, Freeman, San Francisco, 1974.

Rona, P. A., "Plate Tectonics and Mineral Resources," *Sci. Amer.*, 86–95 (July 1973).

Spencer, E. W., *Introduction to the Structure of the Earth*, McGraw-Hill, New York, 1969.

Taylor, H. P., "Oxygen Isotope Evidence for Large-Scale Interaction Between Meteoric Ground Waters and Tertiary Granodiorite Intrusions, Western Cascade Range, Oregon," *J. Geophys. Res.*, **76**, 7855–7873 (1971).

Taylor, H. P. Jr., "The Application of Oxygen and Hydrogen Isotope Studies to Problems of Hydrothermal Alteration and Ore Disposition," *Econ. Geol.*, **69**, 843–883 (1974).

Thompson, G. A., and M. Talwani, "Geology of the Crust and Mantle, Western United States," *Science*, **146**, 1539–1549 (1964).

Verhoogen, J., et al., *The Earth, An Introduction to Physical Geology*, Holt, Reinhart, and Winston, New York, 1970.

White, D. E., *Geothermal Energy -Resources, Production, Stimulation*, Stanford Univ. Press, Stanford, Calif., 1973.

White, D. E., and D. L. Williams, Eds., *Assessment of Geothermal Resources of the United States*, Geological Survey Circular 726, National Center, Reston, Va., 1975.

Table 1.1

Axtel, L. H., "Geothermal Energy Utilization in Japan," *Geotherm. Energy*, **3**, 7 (1975).

Berman, E. R., *Geothermal Energy*, Noyes Data Corp., Park Ridge, N.J., 1975.

Birsic, R. J., *The Geothermal Steam Story*, R. J. Birsic, 1487 Skyline Drive, Fullerton, Calif., 1974.

Guiza, J. L., "Power Generation at Cerro Prieto Geothermal Field," Second U.N. Geothermal Symposium, San Francisco, May 1975.

Kruger, P., and C. Otte, *Geothermal Energy*, Stanford Univ. Press, Palo Alto, Ca., 1973.

Langquist, R., and A. Hansen, "Geothermal Steam Piping at Big Geysers California," Second U.N. Geothermal Symposium, San Francisco, May 1975.

McCabe, B. C., "Practical Aspects of a Viable Geothermal Energy Program," Second United Nations Symposium on the Development and Use of Geothermal Resources, San Francisco, May 1975.

Molina, B. R., and C. J. Banwell, "Chemical Studies in Mexican Geothermal Fields," *Geothermics*, Special Issue, 2, **2** (2), 1385 (1970).

Ragnars, K., et al., "Development of the Namafjall Area, Northern Iceland," *Geothermics*, Special Issue 2, **2** (1), 925 (1970).

Sholes, W. A., "New Zealand Presses Geothermal Power Search," *Geotherm. Energy*, **3**, 42 (1975).

Smith, J. H., and G. R. McKenzie, "Wairakei Power Station New Zealand—Economic Factors of Development and Operating," *Geothermics*, Special Issue, 2, **2** (2), 1717 (1970).

Villa, F. P., "Geothermal Plants in Italy: Their Evolution and Problems," Second U.N. Geothermal Symposium, San Francisco, May 1975.

Table 1.2

Boldizsar, T., "Geothermal Energy Production from Porous Sediments in Hungary," *Geothermics*, Special Issue 2, **2** (1), 99 (1970).

Burrows, W., "Utilization of Geothermal Energy in Rotorua, New Zealand," International Conference on Multipurpose Use of Geothermal Energy, Oregon Institute of Technology, Klamath Falls, Oregon, October 7, 1974.

Dorov, I. M., "Study and Utilization of the Earth's Thermal Energy in the U.S.S.R.," International Conference on Multipurpose Use of Geothermal Energy, Oregon Institute of Technology, Klamath Falls, Oregon, October 7, 1974.

Dragone, G., and O. Rumi, "Pilot Greenhouse for the Utilization of Low Temperature Waters," *Geothermics*, Special Issue 2, **2** (1), 918 (1970).

Einarsson, S. S., "Geothermal Space Heating and Cooling," Second United Nations Symposium on the Development and use of Geothermal Resources, San Francisco, May 1975.

Komagata, S. et al., "The Status of Geothermal Utilization in Japan," *Geothermics*, Special Issue 2, **2** (1), 185 (1970).

Linton, A. M., "Innovative Geothermal Uses in Agriculture," International Conference on Multipurpose Use of Geothermal Energy, Oregon Institute of Technology, Klamath Falls, Oregon, October 7, 1974.

Lund, J., "Utilization of Geothermal Energy in Klamath Falls," International Conference on Multipurpose Use of Geothermal Energy, Oregon Institute of Technology, Klamath Falls, Oregon, October 7, 1974.

Peterson, R. E., and N. El-Ramly, "The Worldwide Electric and Nonelectric Geothermal Industry," *Geotherm. Energy*, **3**, 4–13 (November 1975).

Tikhanov, A. N., and I. M. Dvorov, "Development and Research and Utilization of Geothermal Resources in the USSR," *Geothermics*, Special Issue 2, **2** (2), 1079 (1970).

Wilson, R. D., "Use of Geothermal Energy at Tasman Pulp and Paper Company Limited—New Zealand," International Conference on Multipurpose Use of Geothermal Energy, Oregon Institute of Technology, Klamath Falls, Oregon, October 7, 1974.

CHAPTER 2

Geothermal Brines and Their Chemical and Physical Properties

Currently geothermal energy is tapped by drilling wells into hydrothermal reservoirs consisting of meteoric waters that have been heated from magmatic intrusions. Hydrothermal systems may be heated by convection of water through the magma, by conduction to the convecting currents, by convective mass transfer particularly in geologically active regions, or by combinations of these. As this water circulates through the rock structure by thermal convection, it dissolves and leaches the minerals to form a hot brine solution. In this way, some of the constituents become concentrated. Laboratory experiments in which water has reached equilibrium with rocks at the pressure and temperature of geothermal reservoirs show that the concentration of the constituents are about the same as those found in geothermal brines.

Because circulating water currents in the crust of the earth are both geothermal and hydrothermal systems, there might be some confusion as to the best terminology. Since they are thermal sources within the earth, they are geothermal in nature. On the other hand, since they are thermal waters that have derived their heat from other geothermal sources, "hydrothermal" might be a preferred term. In any case, thermal water, more correctly thermal brines flowing from wells to tap geothermal energy, are from a geologic source, and their thermal energy is derived from geothermal sources. Consequently, the term "geothermal brines" is used throughout this text rather than "hydrothermal brines." This term has the added advantage of being more readily understood in communications involving different disciplines.

Lifetimes of some naturally occurring hydrothermal systems have been estimated between 10,000 and 500,000 yr, depending on the system. Over such a long period of time, changes in the structure of the crust may occur causing changes in chemical constituents leaching into and/or depositing out of the geothermal brines. The natural period of circulation of the systems may exceed 50 yr. Current exploitation of hydrothermal reservoirs is accomplished with flow rates of about 10,000 gal/min from a square mile area. Such a flow rate would correspond to the consumption of 75 ft of water/yr under the square mile area assuming a 30% porosity zone. Thus the withdrawal of geothermal brines from hydrothermal reservoirs over a 10- or 20-yr period at typical flow rates can be visualized as the withdrawal of the hot water from a static natural system with or without replacement. If the original brine is replaced with fresh water, and therefore unsaturated salt content with respect to the heat-producing strata, the rock will be leached with the result that the heat producing strata will become depleted in certain minerals. Thus over a period of time, the brine composition will be altered and the concentrations of some constituents will be reduced. The effect of withdrawal of reservoir fluids on brine composition is demonstrated by the variation of temperature and chloride ion concentration for Broadlands' geothermal brines as shown in Figure 2.1. Variation of the temperature and chloride ion concentration with time and well operating conditions is shown for an East Mesa well of Imperial Valley, California, in Figure 2.2. Chloride ion concentration from well discharges may also decrease because of dilution of the reservoir brine by lower chloride ion content waters from other sources.

The withdrawal of hot fluid from hydrothermal reservoirs results in the flow of steam, water, or a mixture of steam and water. If the well is flowing at a sufficiently high rate, some of the brine will flash to steam because of pressure drop, and a mixed flow of steam and water will result. In principle, it is possible to withdraw the fluid at a high enough rate so that the flashing occurs in the reservoir resulting in steam as well as brine flowing up the well. The production of steam

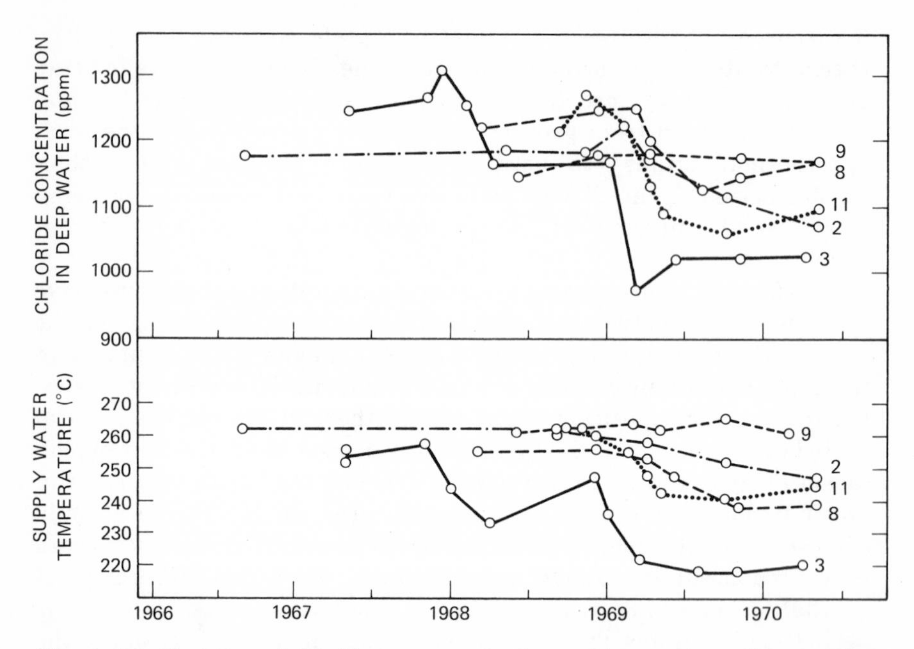

Figure 2.1 Variation of chloride ion concentration and temperature with operating time for the Broadlands, New Zealand, geothermal field. From Mahon and Finlayson (1972).

only from the reservoir is unusual because separation of steam and brine is not readily achieved in most formations. The geysers in California is a rare example of steam production directly from the reservoir.

CHEMICAL COMPOSITION OF BRINES

When geothermal brine flashes to form steam, insoluble gases are released with the first small portion of steam that is flashed. The chemical composition of geothermal brines is conveniently tabulated and studied as the composition of the components in the water system before flashing and the composition of the non-condensible gases after flashing. The total dissolved solids of geothermal brines varies from that of ordinary well water up to concentrated solutions as high as 40% by weight. The brines are mainly sodium chloride solutions. In addition, potassium is found in substantial quantities, being about one tenth that of sodium on an atomic basis. Calcium is the other major

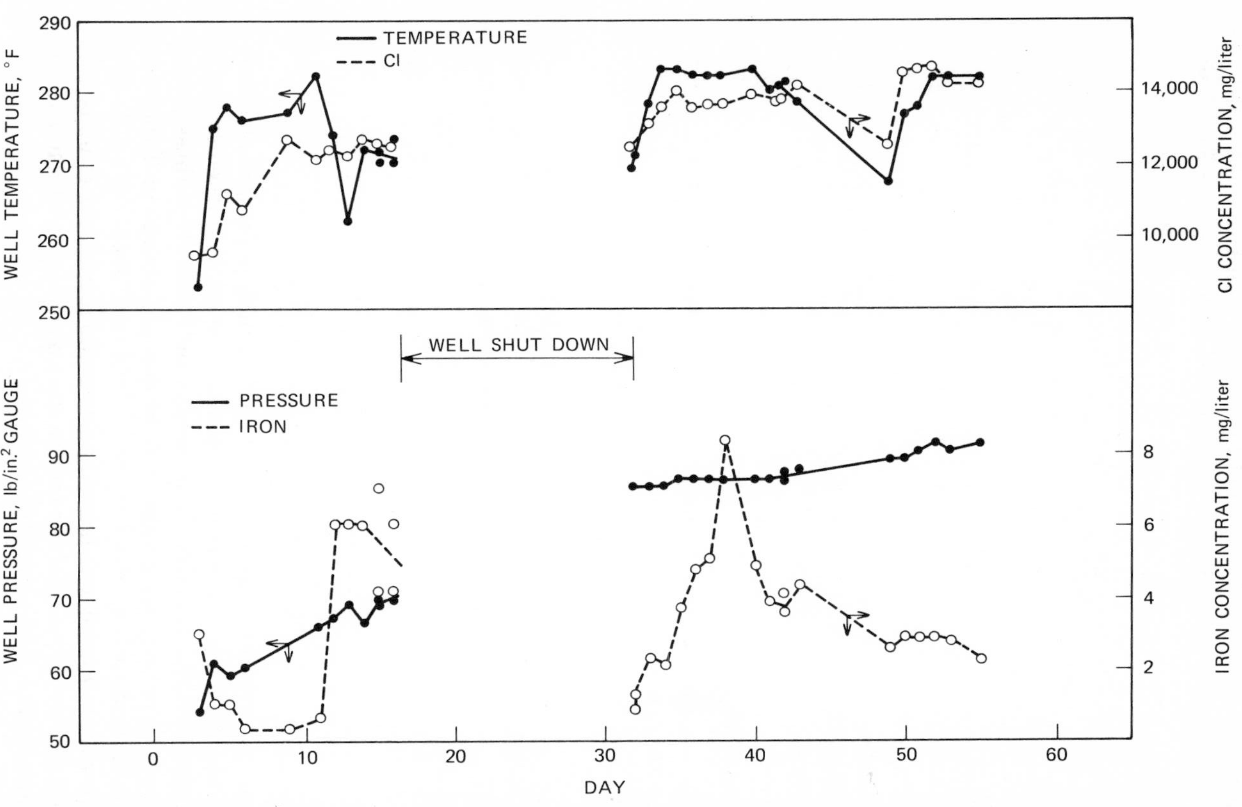

Figure 2.2 Variation of chloride ion concentration, temperature, pressure, and iron concentration over short operating time showing the effect of startup and temporary shutdown of U.S. Bureau of Reclamation well 6-1 at East Mesa, California (Wahl, 1974).

Table 2.1 Distribution of Major Constituent Concentrations of Geothermal Brines Expressed as Weight Percent of Total Dissolved Solids

		Constituent (weight percent)				
Country, Well	Total (ppm)	NaCl	KCl	$CaCl_2$	H_2SiO_3	Total
NZ, Waitapu #6	3,019	72.4	9.7	0.9	20.2	103
NZ, Kawerau 7A	3,375	68.9	8.5	0.28	29.3	107
Jap., Otake #9	3,810	62.6	0.50	1.5	22.8	87
NZ, Broadlands #11	4,003	64.7	10.4	0.05	26.1	101
NZ, Wairakei #44	4,551	73.6	10.7	1.0	14.5	100
Jap., Hatchobaru	5,327	66.5	6.8	0.52	33.6	107
El Sal., Ahuachpan	15,024	81.8	11.1	6.2	4.0	103
Mex., Cerro Prieto M-6	15,637	91.3	6.6	6.4	2.3	107
Mex., Cerro Prieto M-11	20,074	77.4	8.5	4.7	4.3	96
USA, East Mesa 6-1	24,800	72.2	6.8	8.6	1.5	89
Iceland, Reykjanes #8	29,737	72.2	8.1	15.4	2.0	98
USA, Salton Sea I.I.D. #1	258,000	49.6	12.9	30.0	0.2	93
Average	32,279	71.1	8.4	6.3	--	--

cationic constituent of geothermal brines. Silica is present as silicic acid. Chloride ion is the only major anionic constituent, the next most important being bicarbonate. The weight distributions of these major constituents in various geothermal brines are shown in Table 2.1. Sodium chloride is typically 70 to 80% of the total dissolved solids. The data on which this table is based appear to have an error of up to 7%, as indicated by the total of the four major constituents which must be less than 100%.

In general, silica concentration is determined by the equilibrium solubility of quartz and so is a function of the temperature of the geothermal brine in the reservoir. Consequently, the silica concentration in geothermal brines is usually between 200 and 600 parts per million and is independent of the total sodium chloride concentration. In low salinity brines it may be 25% or more of the total dissolved solids, but in higher salinity brines it is less than 5%. Thus the ratio of silicic acid to sodium chloride for the total dissolved solids in geothermal brines is not a constant. Except for silica, the ratio of the major cations, as well as many of the minor ones, is in a constant ratio to that of sodium because of the equilibria reactions involved in the reservoir system. Some chemical equilibria that may be important in controlling the constituent compositions of geothermal brines are shown in Table 2.2.

The concentration of the various constituents in geothermal brines is shown in Table 2.3 for typical wells of various geothermal reservoirs of the earth. The same data expressed in molecular terms are shown in Table 2.4.

Chloride

The chloride ion concentration varies considerably not only from region to region as shown in Table 2.3 but also within a given region from well to well as shown by the ranges of chloride ion concentrations in Table 2.5. This is also true of the total dissolved solids and many of the individual constituents. As an example, variation in chloride and boron concentrations is shown in Figure 2.3 for the southern lowland geothermal region of Iceland.

pH

The pH of the brines are determined by acid-base equilibria* such as carbonic acid-bicarbonate, boric acid-borate, hydrogen floride–floride, ammonia–ammonium, bisulfate–sulfate, silicic acid–silicate, hydrogen sulfate–bisulfate, and reactions involving alumino–silicates such as those shown in Table 2.2. The pH of undisturbed hydrothermal systems has been shown to be controlled by the equilibrium between the brine and the alumino–silicates in the mineral deposits (Ellis, 1969). In general, the pH of carbonate-type geothermal waters tends to be between 6.0 and 6.5 or close thereto because of the buffering action of the carbonate reactions. As geothermal reservoirs are produced, gases will tend to be drawn from the reservoir if diffusion and

*See Chapter 3 for a discussion of the buffering action of acid-base equilibria.

Table 2.2 Chemical Equilibria Important In Controlling Constituent Compositions of Geothermal Brines

Native copper
$8Cu + 8H^+ + SO_4^{2-} \rightleftharpoons 8Cu^+ + 4H_2O + S^{2-}$

Pyrrhotite
$FeS \rightleftharpoons Fe^{2+} + S^{2-}$

Sphalerite, Wurzite
$ZnS \rightleftharpoons Zn^{2+} + S^{2-}$

Galena
$PbS \rightleftharpoons Pb^{2+} + S^{2-}$

Covellite
$CuS \rightleftharpoons Cu^{2+} + S^{2-}$

Acanthite
$Ag_2S \rightleftharpoons 2Ag^{2+} + S^{2-}$

Chalcocite
$Cu_2S \rightleftharpoons 2Cu^+ + S^{2-}$

Pyrite
$FeS_2 + H_2O \rightleftharpoons Fe^{2+} + 1.75S^{2-} + 0.25SO_4^{2-} + 2H^+$

Chalcopyrite
$CuFeS_2 \rightleftharpoons Cu^{2+} + Fe^{2+} + 2S^{2-}$

Bornite
$Cu_5FeS_4 \rightleftharpoons 4Cu^+ + Cu^{2+} + Fe^{2+} + 4S^{2-}$

Cuprite
$Cu_2O + 2H^+ \rightleftharpoons 2Cu^+ + H_2O$

Magnetite
$Fe_3O_4 + 8H^+ \rightleftharpoons Fe^{2+} + 2Fe^{3+} + 4H_2O$

Anglesite
$PbSO_4 \rightleftharpoons Pb^{2+} + SO_4^{2-}$

Cerrusite
$PbCO_3 \rightleftharpoons Pb^{2+}CO_3^{2-}$

Tenorite
$CuO + 2H^+ \rightleftharpoons Cu^{2+} + H_2O$

Smithsonite
$ZnCO_3 \rightleftharpoons Zn^{2+} + CO_3^{2-}$

Siderite
$FeCO_3 \rightleftharpoons Fe^{2+} + CO_3^{2-}$

Dolomite
$CaMg(CO_3)_2 \rightleftharpoons Ca^{2+} + Mg^{2+} + 2CO_3^{2-}$

Anhydrite
$CaSO_4 \rightleftharpoons Ca^{2+} + SO_4^{2-}$

Gypsum
$CaSO_4 \cdot 2H_2O \rightleftharpoons Ca^{2+} + SO_4^{2-} + 2H_2O$

Quartz
$SiO_2 + 2H_2O \rightleftharpoons H_4SiO_4$

Microcline
$KAlSi_3O_8 + 4H^+ + 4H_2O \rightleftharpoons K^+ + 3H_4SiO_4 + Al^{3+}$

Low Albite
$NaAlSi_3O_8 + 4H^+ + 4H_2O \rightleftharpoons Na^+ + 3H_4SiO_4 + Al^{3+}$

Anorthite
$CaAl_2Si_2O_8 + 8H^+ \rightleftharpoons Ca^{2+} + 2Al^{3+} + 2H_4SiO_4$

Hematite

$Fe_2O_3 + 6H^+ \rightleftharpoons 2Fe^{3+} + 3H_2O$

Brucite

$Mg(OH)_2 \rightleftharpoons Mg^{2+} + 2OH^-$

Gibbsite

$Al(OH)_3 \rightleftharpoons Al^{3+} + 3OH^-$

Calcite

$CaCO_3 \rightleftharpoons Ca^{2+} + CO_3^{2-}$

Leucite

$KAlSi_2O_6 + 2H_2O + 4H^+ \rightleftharpoons K^+ + Al^{3+} + 2H_4SiO_4$

Nepheline

$NaAlSiO_4 + 4H^+ \rightleftharpoons Na^+ + Al^{3+} + H_4SiO_4$

Analcite

$NaAlSi_2O_6 \cdot H_2O + H_2O + 4H^+ \rightleftharpoons Na^+ + Al^{3+} + 2H_4SiO_4$

Kaolinite

$Al_2Si_2O_5(OH)_4 + 6H^+ \rightleftharpoons H_2O + 2Al^{3+} + 2H_4SiO_4$

Na-Montmorillonite

$Na_{0.333}Al_{2.333}Si_{3.667}O_{10}(OH)_2 + 7.332H^+ + 2.668H_2O \rightleftharpoons 0.333Na^+ + 2.333Al^{3+} + 3.667H_4SiO_4$

K-Montmorillonite

$K_{0.333}Al_{2.333}Si_{3.667}O_{10}(OH)_2 + 7.332H^+ + 2.668H_2O \rightleftharpoons 0.333Na^+ + 2.333Al^{3+} + 3.667H_4SiO_4$

Ca-Montmorillonite

$Ca_{0.1665}Al_{2.333}Si_{3.667}O_{10}(OH)_2 + 7.332H^+ + 2.668H_2O \rightleftharpoons 0.1665Ca^{2+} + 2.333Al^{3+} + 3.667H_4SiO_4$

Mg-Montmorillonite

$Mg_{0.1665}Al_{2.333}Si_{3.667}O_{10}(OH)_2 + 7.332H^+ + 2.668H_2O \rightleftharpoons 0.1665Mg^{2+} + 2.333Al^{3+} + 3.667H_4SiO_4$

Muscovite

$KAl_3Si_3O_{10}(OH)_2 + 10H^+ \rightleftharpoons K^+ + 3Al^{3+} + 3H_4SiO_4$

Illite

$K_{0.6}Mg_{0.25}Al_{2.30}Si_{3.50}O_{10}(OH)_2 + 8H^+ + 2H_2O \rightleftharpoons 0.6K^+ + 0.25Mg^{2+} + 2.30Al^{3+} + 3.5H_4SiO_4$

Biotite (Annite)

$KFe_3AlSi_3O_{10}(OH)_2 + 10H^+ \rightleftharpoons K^+3Fe^{2+} + Al^{3+} + 3H_4SiO_4$

Source: From Helgeson, 1970, Copyright 1970 by the Mineralogical Society of America.

Table 2.3 Chemical Composition of Typical Geothermal Brines Expressed as Weight Fraction in Parts Per Million of the Total Liquid Brine Arranged in Increasing Chloride Concentration

Country	Well	Temp(°C)	pH	TDS	Cl^-	[b] HCO_3^-	HSO_4^-	S^{2-}	F^-	Br^-	$H_2BO_3^-$	Na^+
Japan	Matsukawa		4.9	3,465[a]	12	37	1,780	-	-	-	11	263
Iceland	Namafjall #5	212	7.5	597	17.7	62.2	37.7	-	0.9	-	-	102.7
Iceland	Reykholtshver	98	9.1	524	90	21.5	72	1.2	2.2	-	0.1	113
Turkey	Denizli-Saraykoy KD-1	195	8.9	5,500[a]	106	2,790	790	-	19.5	-	4.6	1,380
Iceland	Hveragerdi #2	200	7.8	740	125	176	65	-	1.9	-	-	155.5
New Zealand	Rotorua #289	-	8.7	1,060[a]	302	224	39	-	-	1.2	-	312
New Zealand	Orakeikoraku #2	260	9.1	1,781[a]	545	-	142	-	5.7	-	-	550
New Zealand	Waiotapu #6	275	8.9	3,019[a]	1,450	-	52	-	7.5	4.7	-	860
New Zealand	Kawerau #7A	278	6.9	3,375[a]	1,473	-	60	-	1.2	-	-	915
Japan	Otake #9	-	6.7	3,810	1,630	-	145	-	-	-	-	940
New Zealand	Broadlands #11	271	8.2	4,003[a]	1,794	78	10	-	6.4	6.4	8.7	1,020
New Zealand	Wairakei #44	260	8.4	4,551[a]	2,260	-	36	-	8.3	6	-	1,320
Japan	Hatchobaru	250	8.1	5,327[a]	2,327	-	-	-	-	-	1.0	1,396
El Salvador	Ahuchapan	228	7	15,024[a]	8,407	-	-	-	-	-	22	4,839
Mexico	Cerro Prieto M-6	290	6.6	15,637[a]	8,583	134	13	-	-	-	0.9	5,641
Mexico	Cerro Prieto M-11	290	7.9	20,074[a]	11,286	11	-	-	-	-	1.4	6,125
U.S.A	East Mesa 6-1	138	5.8	24,800	14,000	300	173	<1	-	-	-	7,050
Iceland	Reykjanes #8	253	5.7	31,032	17,900	-	141	-	-	6	-	8,450
U.S.A	Salton Sea I.I.D #1	-	5.2	258,000	155,000	108	-	21	-	-	69	50,400

[a] Calculated from individual components

[b] Total CO_2, HCO_3^-, CO_3^{2-} expressed as HCO_3^-

Table 2.3 (Continued)

Well	Ca^{2+}	SiO_2	Mg^{2+}	Li^+	K^+	Fe^{3+}	Rb^{2+}	Cs^+	Other	Source
Matsukawa	23	600	9	-	143	508	-	-	Al^{3+} 29	Nakamura, 1970
Namafjall #5	1.3	318	0.02	-	16.9	-	-	-	-	Thorhallsson, 1975
Reykholtshver	2.6	194	0.03	-	5.7	-	-	-	-	Arnorsson, 1970
Denizli-Saraykoy KD-1	2.4	220	0.35	-	166	-	-	-	-	Dominco, 1970
Hveragerdi #2	1.9	270	0.06	-	12.6	-	-	-	-	Thorhallsson, 1975
Rotorua #289	0.8	180	0.8	-	-	-	0.15	0.25	-	Glover, 1967
Orakeikoraku #2	<1	480	-	3.1	54	-	0.30	0.40	-	Ellis, 1965
Waiotapu #6	10	470	0.06	6.6	155	-	2.4	0.8	-	Ellis, 1965
Kawerau #7	3.5	760	0.16	7.6	152	-	0.85	0.85	-	Ellis, 1965
Otake #9	20	668	110	12.2	10	-	-	-	-	Sato, 1970
Broadlands #11	7.3	805	0.92	-	218	-	1.7	1.4	As^{3+} 5	Mahon, 1972
Wairakei #44	17	660	0.03	14	225	-	2.8	2.5	-	Ellis, 1965
Hatchobaru	10	1,380	-	11	189	-	7	1	-	Koga, 1970
Ahuachapan	341	439	-	-	872	-	-	-	-	Sigvaldason, 1970
Cerro Prieto M-6	366	285	36	13	561	-	-	-	-	Mercado, 1970
Cerro Prieto M-11	340	621	16	17	1,650	-	-	-	-	Mercado, 1970
East Mesa 6-1	770	286	16	54	890	5	-	-	Sr^{2+} 135	Wahl, 1974
Reykjanes #8	1,654	477	-	14	1,260	-	3	3	-	Bjornsson, 1970
Salton Sea I.I.D. #1	28,000	400	54	215	17,500	2,090	137	16	Ba^{2+} 235	Skinner, 1967
									Mn^{2+} 1,560	
									Zn^{2+} 790	
									Pb^{2+} 84	
									Cu^{2+} 8	

Table 2.4 Chemical Composition of Typical Geothermal Brines Expressed As Molal Concentration Millimoles Per Kilogram of the Total Liquid Brine Arranged in Increasing Chloride Concentration

Country	Well	Temp (°C)	pH	Cl^-	$^aHCO_3^-$	HSO_4^-	S^{2-}	F^-	Br^-	$H_2BO_3^-$	Na^+
Japan	Matsukawa		4.9	0.34	0.61	18.5	--	--	--	0.2	11.4
Iceland	Namafjall #5	212	7.5	0.50	1.01	0.39	--	0.05	--	--	4.4
Iceland	Reykholtshver	98	9.1	2.54	0.35	0.75	0.04	0.12	--	--	4.9
Turkey	Denizli-Saraykoy KD-1	195	8.9	2.99	45.7	8.2	--	1.02	--	0.08	60.0
Iceland	Hveragerdi #2	200	7.8	3.52	2.9	0.67	--	0.10	--	--	6.8
New Zealand	Rotorua #289	--	8.7	8.51	3.67	0.41	--	--	0.02	--	13.6
New Zealand	Orakeikoraku #2	260	9.1	15.37	--	1.47	--	0.30	--	--	23.9
New Zealand	Waitapu #6	275	8.9	40.9	--	0.54	--	0.39	0.06	--	37.4
New Zealand	Kawerau #7A	278	6.9	41.6	--	0.63	--	0.06	--	--	39.8
Japan	Otake #9	--	6.7	46.0	--	1.51	--	--	--	--	40.9
New Zealand	Broadlands #11	271	8.2	50.6	1.3	0.10	--	0.34	0.08	0.14	44.4
New Zealand	Wairakei #44	260	8.4	63.8	--	0.38	--	0.44	0.08	--	57.4
Japan	Hatchobaru	250	8.1	65.6	--	--	--	--	--	0.02	60.7
El Salvador	Ahuachapan	228	7	237.2	--	--	--	--	--	0.36	210.5
Mexico	Cerro Prieto M-6	290	6.6	242.1	2.2	0.14	--	--	--	0.01	245.5
Mexico	Cerro Prieto M-11	290	7.9	318.4	0.18	--	--	--	--	0.02	266.5
U.S.A.	East Mesa 6-1	138	5.8	394.9	4.9	1.80	--	--	--	--	306.8
Iceland	Reykjanes #8	253	5.7	504.9	--	1.45	--	--	0.08	--	367.4
U.S.A.	Salton Sea I.I.D.#1	--	5.2	4,372.3	--	--	--	--	--	1.10	2,193.2

[a] Total CO_2, CO_3^{2-}, HCO_3^- expressed as HCO_3^-

Table 2.4 (Continued)

Well	Ca^{2+}	SiO_2	Mg^{2+}	Li^+	K^+	Fe^{3+}	Rb^{2+}	Cs^+	Other		$\Sigma m_i z_{i-}$	$\Sigma m_i z_{i+}$
Matsukawa	0.57	10.0	0.37	--	3.7	9.1	--	--	Al^{3+}	1.07	47	20
Namafjall #5	0.03	5.3	--	--	0.43	--	--	--	--	--	5	2
Reykholtshver	--	3.2	--	--	0.15	--	--	--	--	--	5	4
Denizli-Saraykoy KD-1	0.06	3.7	0.01	--	4.25	--	--	--	--	--	64	58
Hveragerdi #2	0.05	4.5	--	--	0.32	--	--	--	--	--	7	7
Rotorua #289	0.02	3.0	0.03	--	--	--	0.01	0.01	--	--	14	13
Orakeikoraku #2	--	8.0	--	0.44	1.4	--	0.01	0.01	--	--	25	17
Waitapu #6	0.25	7.8	0.01	0.95	3.96	--	0.03	0.01	--	--	43	42
Kawerau #7A	0.09	12.7	0.01	1.09	3.88	--	0.01	0.01	--	--	45	42
Otake #9	0.50	11.1	4.52	1.75	0.25	--	--	--	--	--	53	48
Broadlands #11	0.18	13.4	0.04	--	5.58	--	0.02	0.01	As^{3+}	0.07	51	53
Wairakei #44	0.42	11.0	0.01	2.0	5.75	--	0.03	0.02	--	--	66	65
Hatchobaru	0.25	23.0	--	1.58	4.83	--	0.08	0.01	--	--	68	66
Ahuachapan	8.50	7.3	--	--	22.3	--	--	--	--	--	250	237
Cerro Prieto M-6	9.12	4.8	1.48	1.87	14.3	--	--	--	--	--	273	242
Cerro Prieto M-11	8.47	10.4	0.65	2.44	42.2	--	--	--	--	--	329	319
East Mesa 6-1	19.20	4.8	0.65	7.78	22.8	0.09	--	--	Sr^{2+}	1.54	385	397
Reykjanes #8	41.4	--	--	2.0	32.2	--	0.04	0.02	--	--	484	507
Salton Sea I.I.D.#1	698.3	6.7	2.22	31.0	448	--	1.6	0.12	Ba^{2+}	1.70	4,175	4,373
									Mn^{2+}	28.3		
									Zn^{2+}	12.0		
									Pb^{2+}	0.4		
									Cu^{2+}	0.13		

Table 2.5 Range of Chloride Concentrations from Different Wells in a Given Hydrothermal Reservoir System

Reservoir System	Chloride Ion (ppm) Low	Mean	High
Tokaanu-Waiki, NZ	418	1500	3255
Rotorua, Taupo Volcanic Zone, NZ	100	500	1048
Broadlands, Taupo Volcanic Zone, NZ	1142	1700	1985
Southern Lowlands, Iceland	15	80	272
Denizli-Saraykoy Region, Turkey	72	107	125
Salton Sea Region, USA	10,000	50,000	78,000

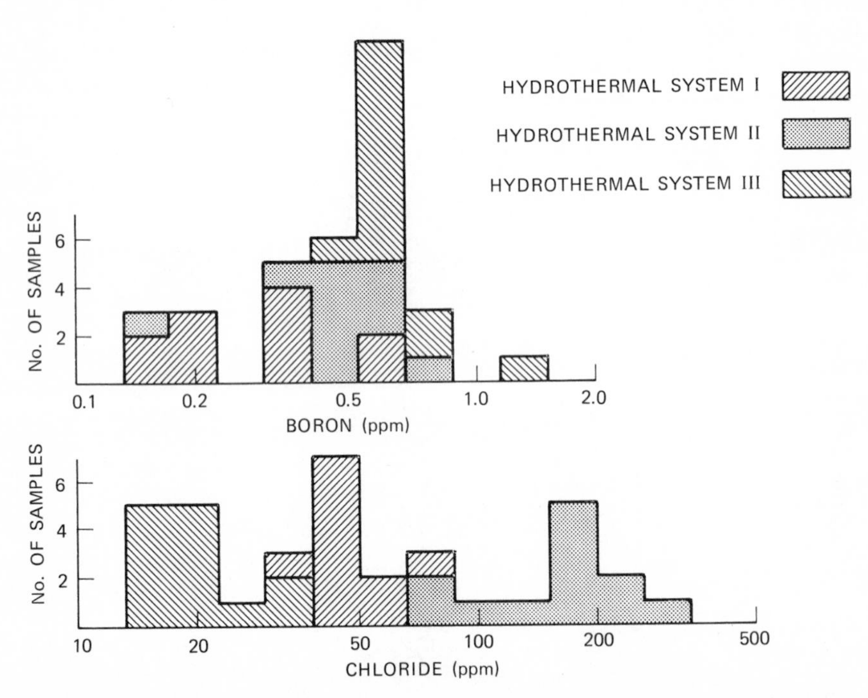

Figure 2.3 Chloride and boron concentrations in three distinct hydrothermal systems within the Southern Lowland, Iceland, geothermal region. From Arnorsson (1970b).

vaporization are possible and variations in pH may result. The reservoir system at Wairakie has shown an increase in pH from 6.3 to 7.0 over 8 yr of operation, probably due to the release of CO_2.

The relationship between the total salinity, which is essentially the sum of the sodium and potassium present, and pH is shown in Figure 2.4. This relationship shows that more saline brines will have a lower pH. For example, the highly saline Salton Sea geothermal brines have a lower pH than the more dilute Icelandic geothermal brines.

In sulfate type brines, the pH is more acidic because of the reactions involved in the formation of sulfate and hydrogen sulfide from sulphur as discussed in the sulfate section to follow.

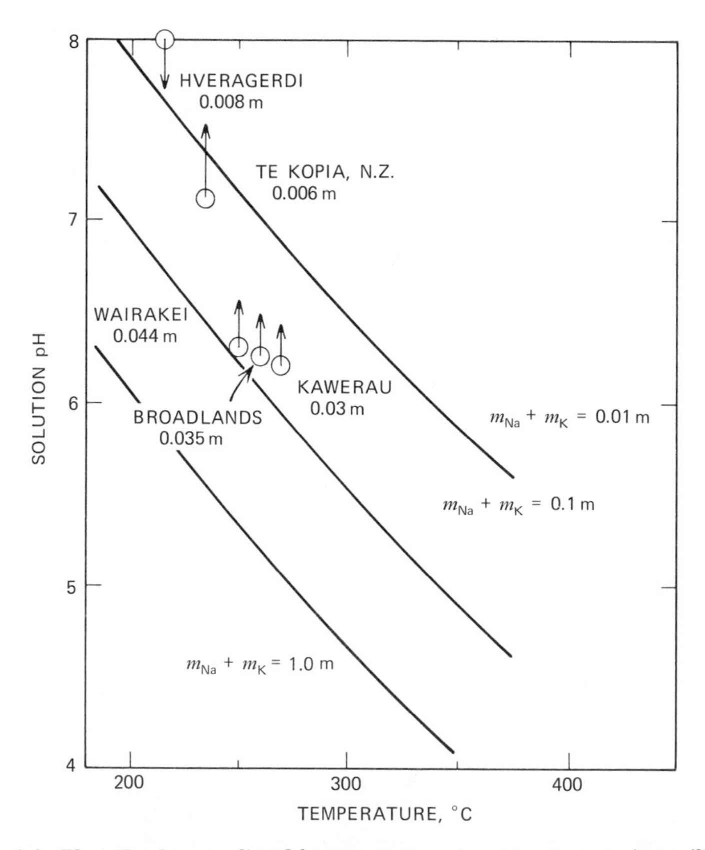

Figure 2.4 The pH values predicted for waters at various temperatures in equilibrium with albite, *K*-mica, and *K*-feldspar. Lines for $(m_{Na} + m_K)$ of 0.01, 0.1, and 1.0 are given. Points show pH values calculated from acid–base equilibria for deep waters in various hydrothermal areas, while arrowheads show the position expected from the mineral equilibrium prediction. From Ellis (1970).

Carbonate and Borate

The ratio of chloride to boron has been shown to be a constant for a given geothermal field, probably because it is determined by equilibria for a given type of rock (Ellis, 1970). Boron is present as borate, which, as shown in Table 2.3, is between 0 and a maximum of about 40 ppm.

The relationship between chloride, borate, and carbonate for various fields is shown in Figure 2.5. Figure 2.6 shows the variation in this ratio for a given field. It will be noted from these figures that the borate is constant although the bicarbonate varies. This is used as a method of determining the source of underground water in a given region (Ellis, 1970; Glover, 1967).

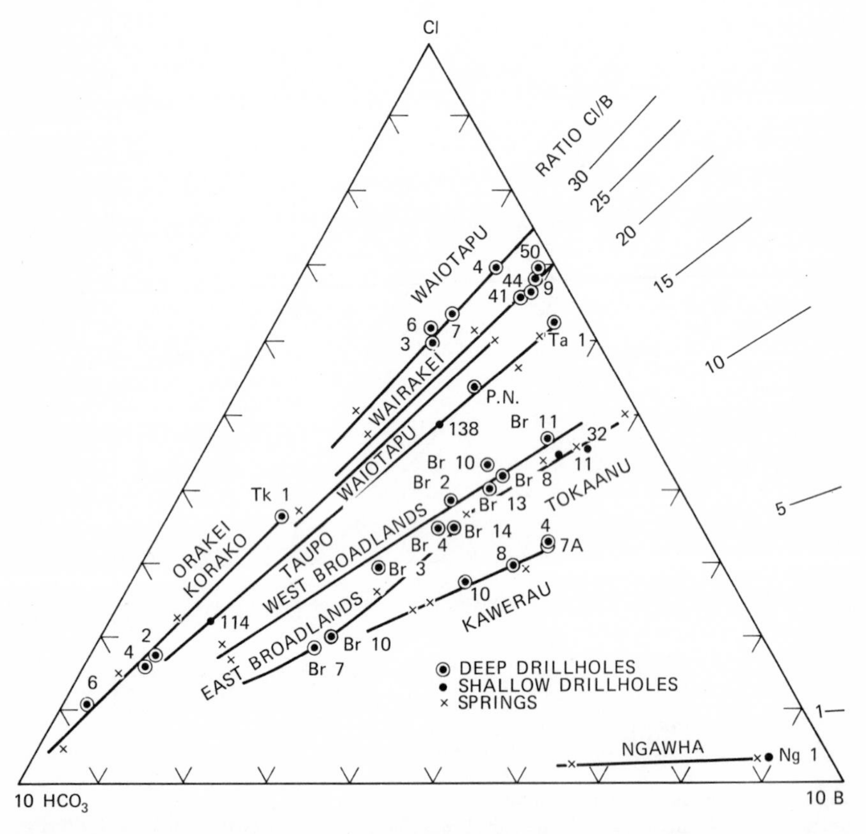

Figure 2.5 Relationship between bicarbonate, borate, and chloride expressed as molecular proportions showing constant chloride to borate ratios for a given hydrothermal system. From Ellis (1970).

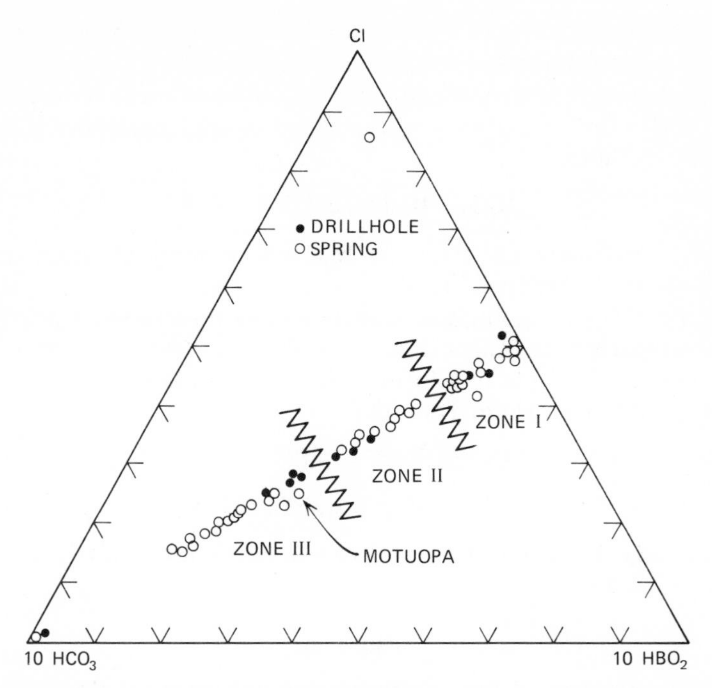

Figure 2.6 Molecular proportions showing variation between identified zones of the Tokaanu-Waihi, New Zealand, hydrothermal region. Adapted from Mahon and Klyen (1968): *New Zealand Journal of Science*, Vol. 11, p. 149, Fig. 3.

The bicarbonate concentration in geothermal brines seldom rises above several hundred parts per million. The amount of carbonate found in the brines will be the result of equilibria with calcium through the calcite reaction, Table 2.2, as well as the pH. As the pH increases the amount of carbonate present will tend to be reduced because of the shift in the bicarbonate–carbon dioxide equilibria toward CO_2.

Sulfate and Sulfide

The presence of sulfur deposits and rock through which the hot water is permeating will give rise to the formation of hydrogen sulfide and sulfuric acid through the reaction:

$$4S + 4H_2O \rightarrow H_2SO_4 + 3H_2S \quad (2.1)$$

The acid developed from this reaction will overide the buffering reactions with the alumino-silicates. If the sulfur deposits are localized in a given region, it may be possible as in Rotokaua, New Zealand, to drill wells deeper and reach sulfur-free water.

In geologically active volcanic regions, sulfur dioxide if present also gives rise to the formation of sulfuric acid through reaction with warm water as follows:

$$4SO_2 + 4H_2O \rightarrow 3H_2SO_4 + H_2S \tag{2.2}$$

Again, if sufficient quantities of SO_2 are present this solution will dominate in controlling the pH.

In both of the foregoing cases, an acid chloride solution is generated, and the bicarbonate concentration is reduced according to the acidity. The amount of sulfate in solution tends to be limited by the solubility of anhydrite or gypsum, Table 2.2:

$$Ca^{2+} + SO_4^{2-} = CaSO_4 \tag{2.3a}$$

$$Ca^{+} + SO_4^{2-} = CaSO_4{\cdot}2H_2O \tag{2.3b}$$

Accordingly, high concentrations of calcium limit sulfate concentrations to low levels.

Fluoride

The concentration of fluoride in geothermal brines is limited by the solubility of calcium fluoride. Fluoride is readily leached from rhyolite, but it is relatively slow from silicified rocks (Glover, 1967). As discussed later in the section on equilibria, the solubility equilibrium relationship, Equations 2.44 or 2.45, rearranged gives the molal concentration of the anion m_X:

$$m_X = \frac{K}{m_{Ca}}$$

Thus both fluoride and sulfate concentrations in brines saturated with these components vary inversely with the calcium concentration.

Sodium

Because of the abundance of sodium chloride and also because of the high solubility of sodium chloride in water, the amount of sodium found in brines is not limited to any particular equilibrium value but depends more on the previous history and temperature of the brines. The sodium content, and, therefore, also the totaled dissolved solids, vary over a large range as shown in Table 2.3. Because of the equilibria between sodium, the alumino-silicates in the rock, and other cations, the ratio of sodium to these cations tends to have particular values as discussed.

Potassium

The concentration of potassium in the brines is determined by the exchange equilibria between sodium and potassium with the alumino-silicates, Table 2.2. Thus the molecular ratio of sodium to potassium is fixed at a given temperature. The variation of the sodium to potassium ratio for water in equilibrium with albite and K-feldspar is shown in Figure 2.7 together with some experimental data. Since chemical equilibria as expressed by the equilibrium constant vary with $1/T$ according to the relation

$$\ln K_{eq} = -\left(\frac{\Delta H}{R}\right)\left(\frac{1}{T}\right) + B \tag{2.4}$$

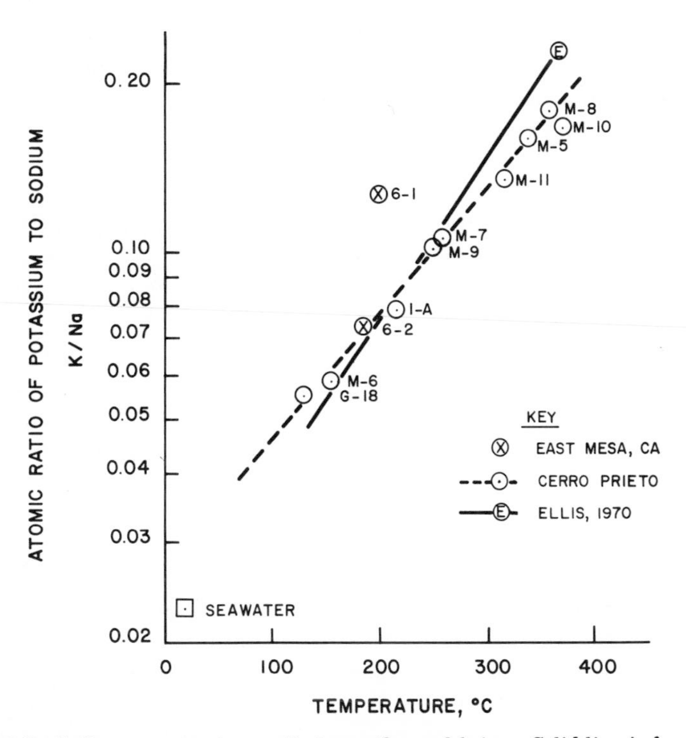

Figure 2.7 Sodium to potassium ratio in geothermal brines. Solid line is for equilibrium conditions with reservoir alumino-silicates (Ellis, 1970). The well numbers are given next to the data point.

the plot is approximately linear on a log versus $1/T$ plot. The sodium to potassium ratio as shown in Figure 2.7 is the basis of the sodium potassium temperature thermometer, that is, the method for predicting the reservoir temperature from a knowledge of the sodium to potassium ratio for brine solutions at wellhead or a surface spring.

Calcium

Because of the equilibria of calcium ion with calcite in the reservoir rock, the product of carbonate and calcium concentrations will be a fixed quantity as a function of temperature. It is strongly dependent on pH, however, which in turn is dependent on the equilibria reactions between the alumino-silicates and sodium in the absence of dominating sulfate chemistry. Thus the product of the calcium and carbonate concentrations will be related to the sodium concentrations and the temperature as shown in Figure 2.8. The relationship applies to the concentration of these constituents for reservoir waters in equilibrium with the rock. As the water flows upward from the reservoir through a well and into process equipment, this relationship no longer applies. Instead the carbonate chemistry, as discussed in the next chapter, controls the equilibria. In general, the carbonate and sulfate concentrations are limited to low levels relative to calcium because of the solubility of $CaCO_3$ and $CaSO_4$.

One rather interesting exception to the low carbonate and sulfate rule with a relatively high calcium concentration is the reservoir in the Denizli–Saraykoy area of Turkey. In that system, Table 2.3, the sulfate concentration is about 1000 ppm, the bicarbonate concentration is 2000 to 3000 ppm, while the calcium concentration is only 2 to 3 ppm. It is worth noting, however, that the product of the calcium and bicarbonate concentration fits the theory relating these concentrations to that of the total sodium concentration as shown in Figure 2.8.

Lithium

Lithium is found in substantial parts per million quantities, that is up to several hundred parts per million. Lithium does not concentrate in the brine solution, so its concentration is similar to that in the rock in which the reservoir exists (Ellis, 1965).

Iron and Aluminum

Iron and aluminum both form insoluble complex salts with silica. Thus as would be expected, both iron and aluminum are absent or in

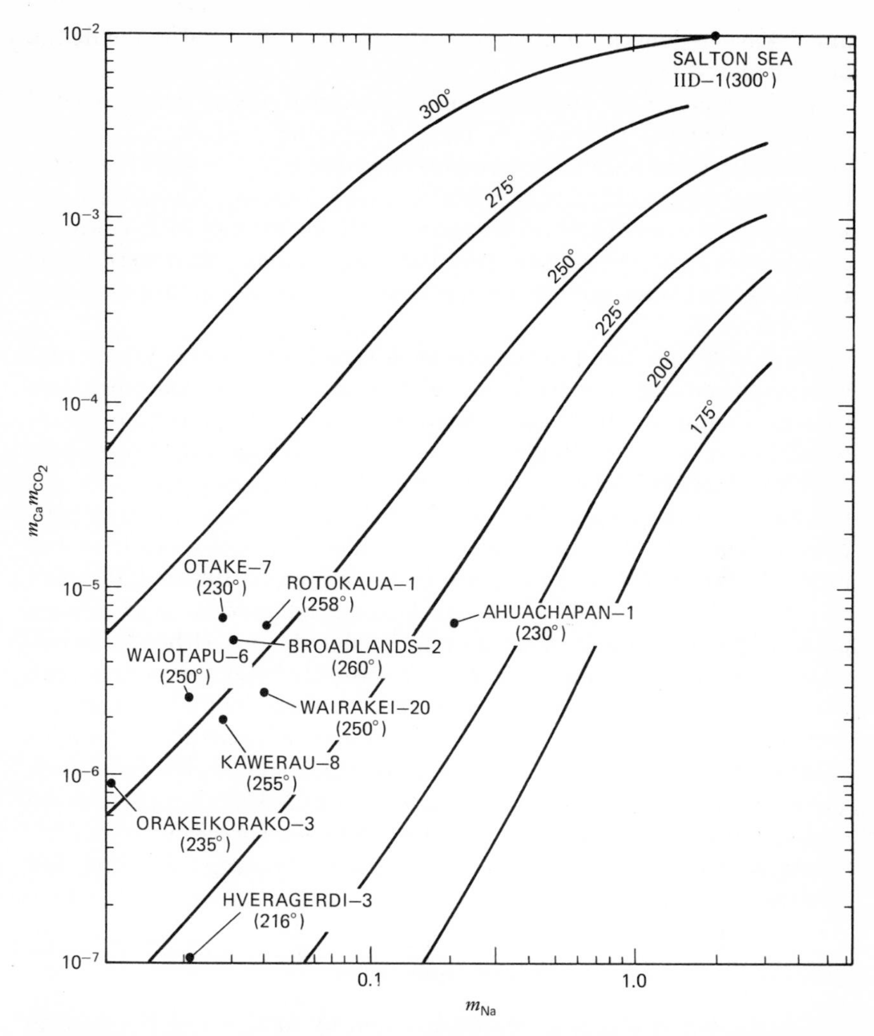

Figure 2.8 Relationship of the product of calcium and carbonate ion concentrations with temperature and sodium ion concentrations for equilibrium reservoir conditions. From Ellis (1970).

minute quantities in geothermal brines. Iron being somewhat more soluble is sometimes found in quantities near the 1-ppm level as shown in Table 2.3.

Trace Metals

Very little data have been compiled on the composition of trace elements in geothermal brines. The data that are given in Tables 2.3

and 2.4 are in general limited to a few measurements and so should not be considered typical.

A rather complete analysis has been carried out on some thermal waters in Iceland (Arnorsson, 1970). Chromium, cobalt, nickel, and zinc are found in only a few samples of acidic water or alkaline water that came from acid clay. Copper was not detected. The cobalt and zinc content was found to increase with decreasing pH. Gallium, germanium, iron, vanadium, titanium, and molybdenum were found in substantial trace quantities in a large number of samples as shown in Figure 2.9.

Strontium was found to be present in East Mesa Well 6-1 brine at a concentration of 100 ppm (Wahl, 1974), which is somewhat more than that of lithium in that brine. Magnesium and manganese are sometimes found in 0.1- to 10-ppm quantities. In high salinity brines over 1% total dissolved solids, such as from the Salton Sea field, they are frequently in the range of 0.1 to 1% of the total solids or 500 to 5000 ppm. Zinc and lead are generally in less than 0.1-ppm quantities, but in the Salton Sea field brines are in the range of 100 to 1000 ppm. Barium, rubidium, and cesium are frequently reported in the range 0.1 to 1 ppm and higher in the Salton Sea field brines. Other elements, when analyses are done for them, are usually reported in less than 0.1-ppm quantities except for the more saline brines in which they are proportionately higher. The reported absence of elements that might be present in less than 10-ppm quantities should be considered questionable unless special care is taken in sampling. This is because such small quantities can easily be lost as minute precipitates or even reactants with sampling container surfaces between sampling and analysis (Wahl, 1974).

Gas Composition

The composition of the gases determined by analysis of the flashed steam from the brine water at the surface is given in Table 2.6. Carbon dioxide is the major gaseous component of most geothermal brines, generally being greater than 90% by volume of the gaseous components. Exceptions would be those of various acidic waters where considerable sulfur or sulfur dioxide is present generating hydrogen sulfide as discussed above. This is exemplified by the Namafjall, Iceland, analysis in Table 2.6.

Ammonia is found in geothermal waters that have come in contact with organic matter. In general, such organic matter is present in very low quantity or absent entirely in the rock structures near magmatic

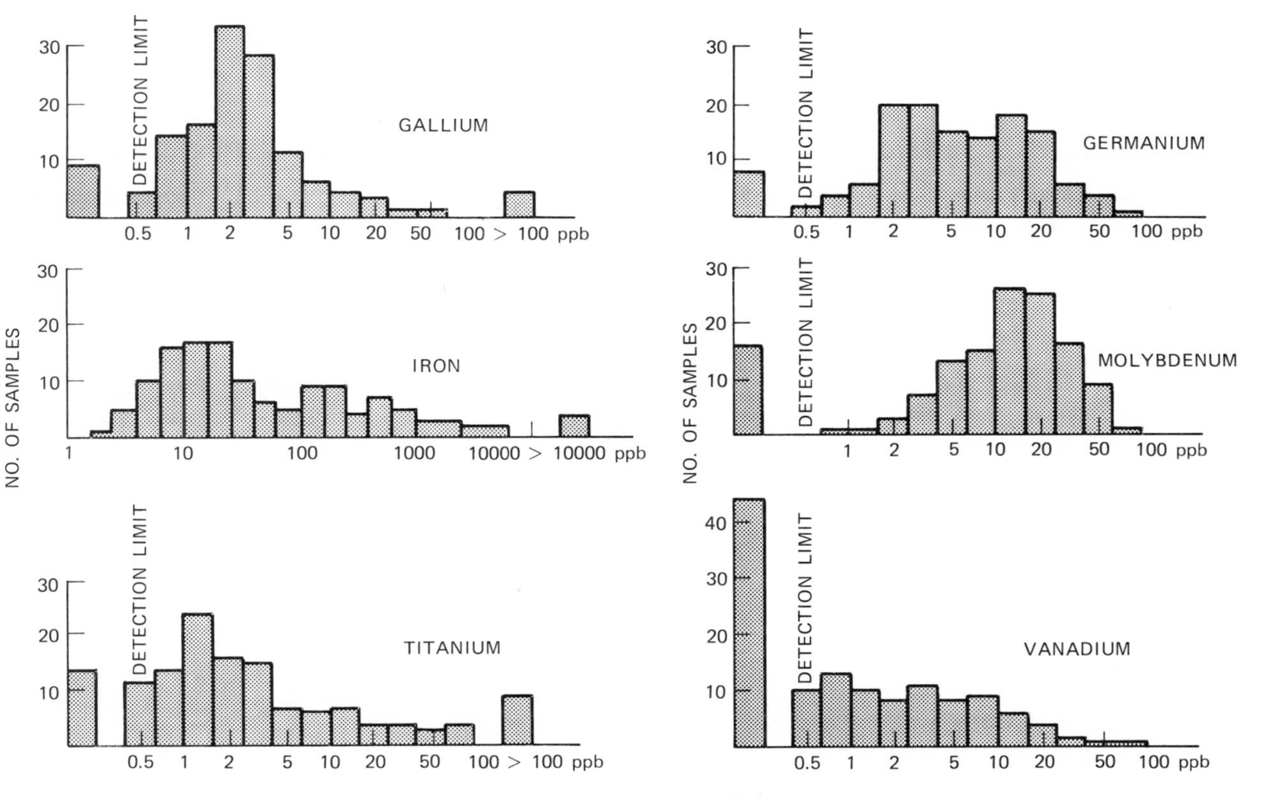

Figure 2.9 Distribution of concentrations of trace elements in different samples from various hydrothermal regions of Iceland. From Arnorsson (1970b).

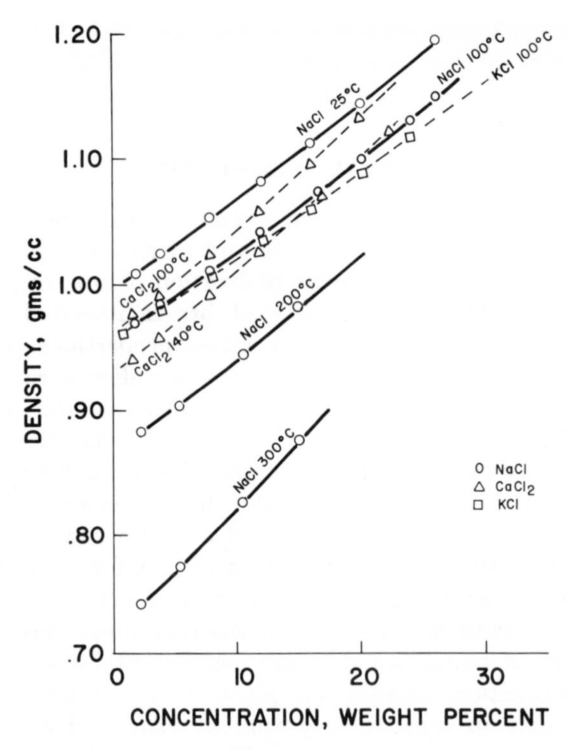

Figure 2.10 Density of aqueous salt solutions at different temperatures for the major constituents of geothermal brines. Data from Perry (1973).

between these ionic constituents in dilute solutions and relatively weak effects at higher concentrations, it is not surprising that the plots in Figure 2.10 of density versus concentration are linear. The slopes of the straight lines at 100°C in Figure 2.10, that is, the change in density per unit change in the concentration expressed in weight percent, are 0.0072 for sodium chloride, 0.0070 for potassium chloride, and 0.0089 for calcium chloride. Experimental data in the range 20 to 200°C, as typified by the NaCl and $CaCl_2$ lines in Figure 2.10, shows that these slopes are independent of temperature. Since geothermal brines are generally 70% or more sodium chloride and since the next major constituent, potassium chloride, has very close to the same effect on density as sodium chloride, the density of geothermal brines can be estimated by correcting the density of water by

$$\rho = \rho_W + 0.0073 w_t \tag{2.5}$$

In this equation, 0.0073 represents the weighted average of the slopes for the "average" brine in Table 2.1 rather than 0.0072 for pure sodium chloride. Comparison with geothermal brines shows Equation 2.5 to be less than 2% in error for brines as concentrated as 3.5wt.%. From Figure 2.11 the density of pure water can be obtained at the temperature of the brine and corrected using Equation 2.5 to determine the density of the brine up to 200°C. Alternatively, the density of water can be determined (Keenan, 1951) by

$$\rho_W = \frac{(1 + dt^{1/3} + et)}{(v_c + at^{1/3} + bt + ct^4)} \tag{2.6}$$

where $v_c = 3.1975$ cm^3/g; $t = 647.11 - T$ or $= 374.11 - T$ for T in °C; $a = -0.3151548$; $b = -1.203374 \times 10^{-3}$; $c = +7.48908 \times 10^{-13}$; $d = +0.1342489$; and $e = -3.946263 \times 10^{-3}$.

Above 200°C, the slope of the density versus concentration line, which is 0.0073 in Equation 2.5, is not constant. As shown in Figure 2.10, the slope for NaCl at 200°C is 0.0079 and at 300°C is 0.0107. Thus the density of brine above 200°C as given by Equation 2.5 should use

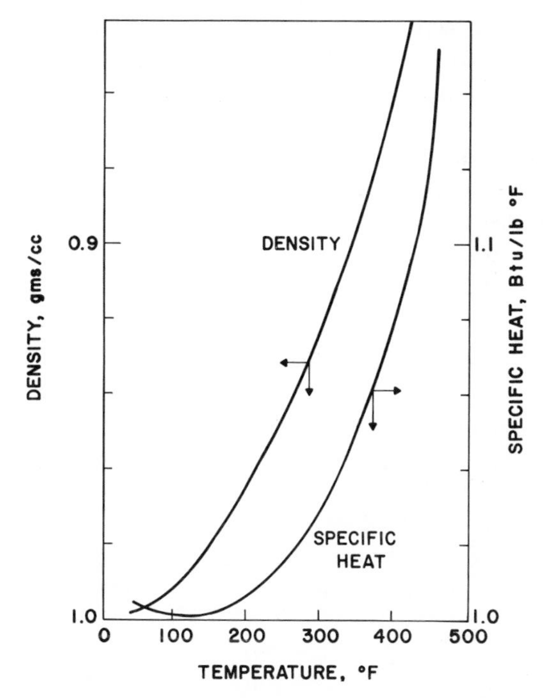

Figure 2.11 Density and specific heat of water as a function of temperature. Data from Keenan and Keyes (1951).

an appropriately corrected slope:

$$\rho = \rho_W + 0.0073[1 + 1.6 \times 10^{-6}(T - 273)^2]w_t \tag{2.7}$$

Above a concentration of 20wt.%, that is, 200,000 ppm, and temperature above 200°C the slope of density versus concentration line begins to fall off (Haas, 1970). For a more accurate estimate for these conditions, the compilation by Haas (1970) may be used.

The presence of dissolved carbon dioxide, the major gaseous constituent of geothermal brines, does not affect the density except near the critical point. The presence of CO_2 lowers the critical temperature of the solution, thus substantially affecting the properties of the solution near this region.

Heat Capacity

The heat capacity of solid salts such as sodium chloride is 0.2 compared to water, which is 1. In aqueous solutions, the ionized salt atoms are widely separated and each individually surrounded by water molecules. Because of the large polar forces between water molecules, the energy required to increase the velocity of the water molecules and therefore its heat capacity is large compared to that necessary to increase the velocity of the individual ionized salt atoms. For these reasons, the specific heat of aqueous solutions is often estimated by assuming the heat capacity of the salt to be negligible. Thus a brine containing 10wt.% salts would have a specific heat of 0.9, while a 20wt.% solution would have a specific heat of 0.8. Accordingly,

$$c = c_W\left(1 - \frac{w_t}{100}\right) \tag{2.8}$$

Using this equation, brine from East Mesa well 6-1, which according to Table 2.3 has a total dissolved solids of 2.48%, would have a heat capacity equal to 0.975 times the heat capacity of pure water.

If there were no heat effect on dissolving salts in water, then the heat capacity would be obtained by adding the weighted heat capacity of each of the components to that of water as shown in Equation 2.9:

$$c = c_W\left(1 - \frac{w_t}{100}\right) + \sum_i \frac{c_i w_i}{100} \tag{2.9}$$

The heat capacities of the major constituents are given by the set of

Table 2.7 Heat Capacity (Btu/lb°F) of Pure Crystalline Salts at Various Temperatures

Crystalline Salt	Temperature (°C)		
	0	100	200
NaCl	0.206	0.213	0.220
KCl	0.159	0.164	0.170
$CaCl_2$	0.161	0.164	0.168
Mixture of 10 NaCl: 1 KCl	0.202	0.209	0.215

Equations 2.10:

$$\text{NaCl:} \quad c_1 = 0.186 + 7.24 \times 10^{-5} T \tag{2.10a}$$

$$\text{KCl:} \quad c_2 = 0.146 + 5.08 \times 10^{-5} T \tag{2.10b}$$

$$\text{CaCl}_2\text{:} \quad c_3 = 0.152 + 3.48 \times 10^{-5} T \tag{2.10c}$$

Since the heat capacity of these solid salts as shown in Table 2.7 is about 0.2 particularly for a 10:1 Na/K ratio, the heat capacity for brine would be estimated from the concentration of total dissolved solids:

$$c = c_W \left(1 - \frac{w_t}{100}\right) + 0.002 w_t \tag{2.11}$$

Because of the variation of the ionic activity of the salts in water, that is, the variation in interactions between the ions and water molecules with temperature, the heat of solution of salts vary with temperature. Sodium chloride, for example, exhibits a minimum heat of solution, that is, a minimum enthalpy decrease on dissolving, at 50°C. Including this effect in the estimation of heat capacity is accomplished by adding a correction term that is negative above 50°C to Equation 2.11 to obtain

$$c = c_W \left(1 - \frac{w_t}{100}\right) + 0.002 w_t + b w_t \tag{2.12}$$

where $b = 1.71 \times 10^{-4} (dH_s/dT)$.

Derivation of Equation 2.12. The enthalpy change for the process of mixing w_t parts of salt and $100 - w_t$ parts water at the temperature T and then heating the resulting salt solution from the temperature T to the temperature $T + dT$ is

$$\frac{w_t H_{S,T}}{58.5} + 100 c dT$$

The enthalpy change for the process of heating w_t parts salt and $100 - w_t$ parts water separately from the temperature T to the temperature $T + dT$ and then mixing the heated salt and water at the temperature $T + dT$ is

$$(100 - w_t) c_W dT + w_t c_s dT + \frac{w_t H_{s,T+dT}}{58.5}$$

Since enthalpy is a state function and both of the preceding processes start and end in the same states, the two enthalpy changes must be equal so that after equating and rearranging

$$c = c_W \left(1 - \frac{w_t}{100}\right) + \frac{w_t c_s}{100} + 1.71 \times 10^{-4} \frac{(H_{s,T+dT} - H_{s,T})}{dT}$$

The last term is simply dH_s/dT so that Equation 2.12 is obtained.

Since the latter term in Equation 2.12 is relatively small, a simple polynomial approximation for dH_s/dT as a function of temperature can be used for estimating the heat capacity of salt solutions. Experimental data for infinitely dilute solutions (Gardner, 1969; Liu and Lindsay, 1971) show that b has the value

$$b = 0.0062 + 0.00016 \left[\frac{(T - 50)}{100}\right]^3$$

in the range 50 to 300°C. For increasing concentration of salt, this correction factor is smaller. Experimental data, Figure 2.12, shows that the negative correction factor b decreases with increasing salt concentration by the factor $(1 - 0.21 w_t^{0.4})$ so that the correction term for the effect of heat of solution is

$$b w_t = -\left\{0.0062 + 0.0016 \left[\frac{(T - 50)}{100}\right]^3\right\} [1 - 0.21 w_t^{0.4}] w_t \quad (2.13)$$

Inserting this value of b in Equation 2.12 gives

$$c = c_W \left(1 - \frac{w_t}{100}\right) + 0.002 w_t - \left\{0.0062 + 0.0016 \left[\frac{(T - 50)}{100}\right]^3\right\} \times [1 - 0.21 w_t^{0.4}] w_t \quad (2.14)$$

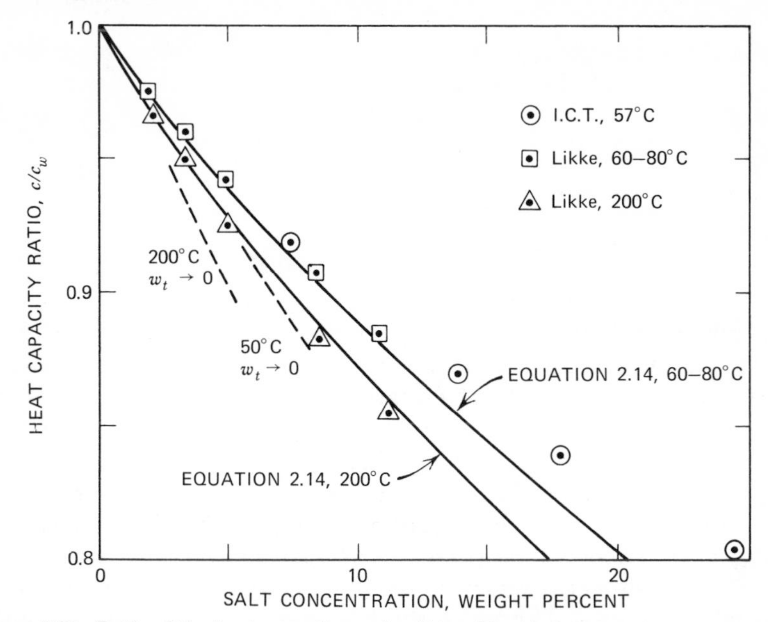

Figure 2.12 Ratio of the heat capacity c of sodium chloride brine to that of pure water c_w as a function of salt concentration. The dashed lines represent the slope of the curve at zero salt concentration from the experimental data of Gardner (1968) and Liu and Lindsay (1971) for 50 and 200°C. Experimental data at 57°C is from the *International Critical Tables* and in the range 60 to 200°C from Likke and Bromley (1973).

Experimental data for heat capacities of sodium and potassium chloride solutions (Likke and Bromley, 1973) shows that both have the same heat capacities so that this equation applies to geothermal brines where w_t is the concentration of total dissolved salts. This is as expected because sodium and potassium chloride have nearly the same heat capacity, and both are small relative to water so that any difference becomes insignificant in calculating the solution heat capacity. Plots of this equation are shown in Figure 2.12 for 68 and 200°C as a function of salt concentration. Since $-dH_s/dT$ is a minimum and zero at 50°C, the curve for heat capacity at this temperature, or about 60°C, is the upper limit for heat capacity of a brine. Thus the variation of heat capacity and concentration for brines is sufficiently well defined by the band between 200 and 60°C for most engineering purposes.

In the range of 100 to 200°C and below 10% salt, which is the principle region of interest for geothermal brines, the correction factor bw_t has the value $-0.004w_t$ so that

$$c = c_W\left(1 - \frac{w_t}{100}\right) - 0.002w_t \qquad (2.15a)$$

Between 10 and 20wt.% salt

$$c = c_W\left(1 - \frac{w_t}{100}\right) \tag{2.15b}$$

and above 20wt.% salt

$$c = c_W\left(1 - \frac{w_t}{100}\right) + 0.001w_t \tag{2.15c}$$

The difference between the heat capacity calculated by Equation 2.14 and Equations 2.8 or 2.15 is not sufficiently large compared with the accuracy and complexity of Equation 2.14 to justify its use for most purposes. Thus for most engineering purposes, either Equation 2.8 or 2.15 will be sufficiently accurate since the correction term to the heat capacity will be less than ±2% of the heat capacity below 200°C.

Another but less direct approach is to estimate the heat capacity using partial molal quantities. The total property of a substance is equal to the sum of the partial molal quantity times the mole fraction of each substance. Therefore the heat capacity per 100 g of sodium chloride solution is given by

$$100c = w_1\left(\frac{\bar{c}_S}{58}\right) + (100 - w_1)\left(\frac{\bar{c}_w}{18}\right) \tag{2.16a}$$

The partial molal heat capacity of sodium chloride at infinite dilution decreases with increasing temperature above 50°C. Since $\bar{c}_W = c_W$ at infinite dilution, Equation 2.16*a* becomes

$$100c = c_W(100 - w_1) + w_1\left(\frac{\bar{c}_S}{58}\right) \qquad \text{for } w_1 \to 0 \tag{2.16b}$$

Comparing this with Equation 2.16*a* shows the effect of heat of solution on the heat capacity. Approximating partial molal heat capacity data (Gardner, 1969; Liu, 1971) with a polynomial and inserting in Equation 2.16*b* results in

$$c = c_W\left(1 - \frac{w_1}{100}\right) - [0.23 + 0.002(T - 273) + 2 \times 10^{-5}(T - 273)^2]\left(\frac{w_1}{100}\right)$$

$$\text{for } w_1 \to 0 \tag{2.16c}$$

This can be modified as indicated to include the effect of experimental data on heat capacity versus concentration similar to Equation 2.14.

Vapor Pressure

The vapor pressure of water over an aqueous salt solution is given by Raoult's law

$$p = P_W x_W \tag{2.17}$$

Alternatively, the decrease in vapor pressure Δp due to a salt of mole fraction x_s is given by rewriting Equation 2.17 in the form

$$\Delta p = P_W x_S \tag{2.18}$$

A monovalent salt that is 100% ionized, such as a dilute solution of sodium chloride in a concentration of 500 mmol/liter will decrease decrease the vapor pressure at 100°C 13.6 mm according to Equation 2.18. The decrease in vapor pressure of water for aqueous solutions of the major constituents of geothermal brines is given in Table 2.8. This table shows that the monovalent salts of sodium chloride, potassium chloride, and others cause a decrease in vapor pressure of 12 to 12.3 mm, which shows that they are close to 100% ionized for the more dilute solutions. Sodium bicarbonate causes a decrease in vapor pressure of 12.9 mm of mercury, which is somewhat higher than that of sodium chloride, perhaps due to the partial ionization of the bicarbonate ion. The divalent salts, such as calcium chloride, cause a decrease in vapor pressure of about 17 mm. Since a divalent chloride salt contains three ions to the two ions of sodium chloride, the expected decrease would be in the ratio of 3:2 to that of sodium chloride or about 18 mm of mercury. Thus this salt is ionized almost to

Table 2.8 Decrease of Vapor Pressure of Water (mm Hg) Due to the Presence of Inorganic Salts at 100°C

	Concentration of Salt (millimoles/liter)								
Substance	500	1000	2000	3000	4000	5000	6000	8000	10,000
NaCl	12.3	25.2	52.1	80.0	111.0	143.0	176.5	--	--
NaBr	12.6	25.9	57.0	89.2	124.2	159.5	197.5	266.0	--
$NaHCO_3$	12.9	24.1	48.2	77.6	102.2	127.8	152.0	198.0	239.4
KCl	12.2	24.4	48.8	74.1	100.9	128.5	152.2	--	--
LiCl	12.1	25.5	57.1	95.0	132.5	175.5	219.5	311.5	393.5
$CaCl_2$	17.0	39.8	95.3	166.6	241.5	319.5	--	--	--
$SrCl_2$	16.8	38.8	91.4	156.8	223.3	281.5	--	--	--
$MgCl_2$	16.8	39.0	100.5	183.3	277.0	377.0	--	--	--

Source. Weast, 1975, Handbook of Chemistry & Physics, page E-1.

the same extent as sodium chloride. The average decrease in vapor pressure per 500 mmol ion present is about 6 mm, or 24 mm/g-mol of monovalent salt. For geothermal brines, which are mostly monovalent chloride salts, a simple way to estimate the vapor pressure lowering at 100°C is to multiply the gram-moles of chloride ion per liter by 24. At any other temperature the decrease in vapor pressure would be obtained by the ratio of the vapor pressure of water at the new temperature to that at 100°C; that is,

$$\Delta p = (24/760) p_W x_{Cl} \tag{2.19}$$

Converting gram-moles to weight percent and noting that $p = P_W - \Delta p$, the vapor pressure of the brine is given by

$$p = P_W \left(1 - \frac{0.009 w_{Cl}}{\rho} \right) \tag{2.20}$$

The data in Table 2.8 shows that the decrease per unit concentration is about constant, being about 10% higher at salt concentrations of 20%. Consequently, Equation 2.20 will give a reliable estimate over all brine concentrations for which chloride is the predominant anion. For typical geothermal brines, the chloride concentration is related to the total dissolved solids so that the above equation becomes

$$p = P_W \left(1 - \frac{0.004 w_t}{\rho} \right) \tag{2.21}$$

For East Mesa well 6-1 this gives $p = 0.990 P_W$, which agrees quite well with Equation 2.22, estimated by the more accurate method described herein.

A more precise estimate of the vapor pressure lowering can be estimated by using the vapor pressure lowering for the major constituents from Table 2.8. For example, the concentration of East Mesa well 6-1 brine as obtained from Table 2.4 for the monovalent cations is 337 mmol/liter, for the divalent cations is 21.4 and for silica is 4.8 giving a total concentration of 363 mmol/liter. From Table 2.8 using data for a concentration of 500 mmol/liter, which is close to a concentration of 363, the decrease in vapor pressure is 12.3 mm for sodium chloride, which is the major monovalent salt; 17.0 mm for calcium chloride, which is the major divalent salt; and would be 6.8 mm for a monomolecular species such as silica. Thus the computation of the decrease in vapor pressure at 100°C by summing the decrease due to each of the constituents is as follows:

$$(0.337)\left(\frac{12.3}{0.5}\right) = 8.290 \text{ mm}$$

$$(0.021)\left(\frac{17.0}{0.5}\right) = 0.714 \text{ mm}$$

$$(0.005)\left(\frac{6.8}{0.5}\right) = \underline{0.068 \text{ mm}}$$

$$\Delta p \ @ \ 100°\text{C} = 9.07 \text{ mm}$$

The vapor pressure decrease at a temperature T is then

$$\Delta p \ @ \ T = \left(\frac{9.07}{760}\right) P_{W,T} = 0.012 P_{W,T}$$

Consequently, the vapor pressure for East Mesa brines at any temperature T is given by

$$p_{B,T} = 0.988 P_{W,T} \tag{2.22}$$

which states that the vapor pressure of the brine is equal to 98.8% of the vapor pressure of pure water at the same temperature.

The vapor pressure of water can be obtained from Figure 2.13 or

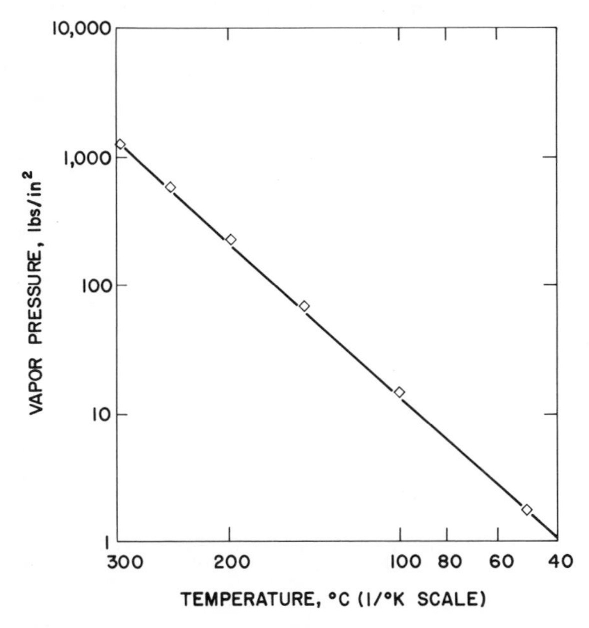

Figure 2.13 Vapor pressure of saturated water. Data from *Handbook of Chemistry and Physics* (Weast, 1975, p. D-181).

calculated by (Keenan and Keyes, 1951)

$$\log_{10}\left(\frac{p_c}{P_W}\right)=\frac{(ax + bx^2 + cx^3 + ex^4)}{T(1+dx)} \tag{2.23}$$

where P_W is the vapor pressure, atm; p_c = 218.167 atm; T is the temperature, °K; $x = 647.27 - T$; for:

	50 to 347.11°C	10 to 150°C
a =	3.3453130	3.2437814
b =	4.14113×10^{-2}	5.86826×10^{-3}
c =	7.515484×10^{-9}	1.1702379×10^{-8}
d =	1.3794481×10^{-2}	2.1878462×10^{-3}
e =	6.56444×10^{-11}	0

Surface Tension and Viscosity

In an ionic solution, the increased electrostatic forces resulting from the ions will increase the forces of attraction on the surface layers of water molecules, thus increasing the surface tension of an ionic salt solution. Similarly, the increased electrostatic forces in the bulk of an ionic solution will increase the transmission of sheer forces throughout the fluid, thus increasing viscosity. Experimental data for the viscosity and surface tension of aqueous salt solutions is limited.

The viscosity of water as shown in Figure 2.14 varies with tempera-

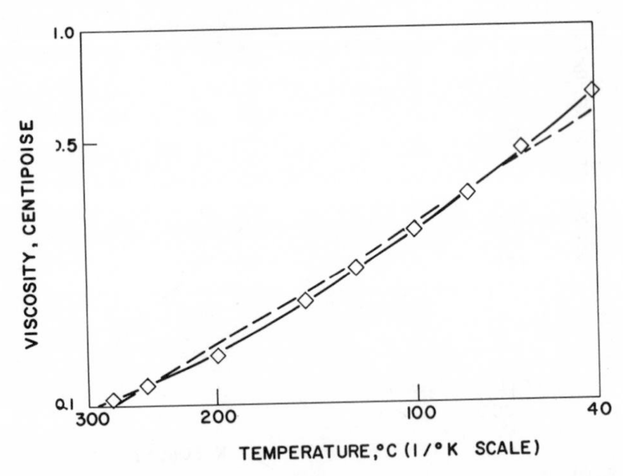

Figure 2.14 Viscosity of water as a function of temperature. The dashed line corresponds to Equation 2.24. Data from Dorsey (1968, p. 184–185).

ture according to the equation:

$$\log \mu_w = -2.03 + \frac{560}{T} \tag{2.24}$$

The effect of aqueous solutions of salts, which are the major constituents of geothermal brines on viscosity, are shown in Figure 2.15 for ordinary temperatures. The equations for each of the curves in this figure are:

$$\text{NaCl: } \frac{\mu}{\mu_W} = 0.022 w_t + 0.00025 w_t^2$$

$$\text{KCl: } \frac{\mu}{\mu_W} = 0.0043 w_t + 0.0001 w_t^2$$

$$\text{CaCl}_2\text{: } \frac{\mu}{\mu_W} = 0.00271 w_t + 0.001 w_t^2$$

Computing the weight average of these curves for a typical geothermal brine gives

$$\mu = \mu_W(1 + 0.021 w_t + 0.00027 w_t^2) \tag{2.25}$$

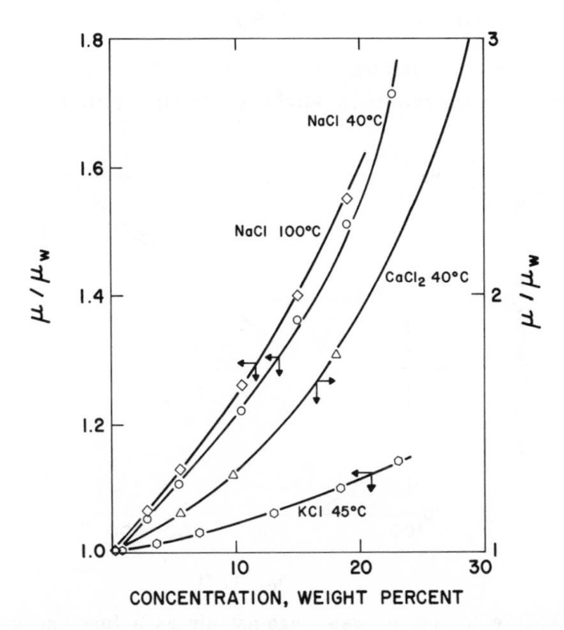

Figure 2.15 Ratio of the viscosity of aqueous salt solutions to that of pure water as a function of concentration. Data from *International Critical Tables* (Vol. 5, p. 14–17).

Note that this is sensitive to abnormally high concentrations of calcium chloride such as are found in the geothermal brines of the Salton Sea region. For such cases, a new weighted average relationship should be obtained.

The surface tension of any solution will approach zero at its critical temperature because the surface tension of a gas is zero. Thus a plot of surface tension versus temperature below the critical temperature yields a straight line on log–log paper. This is shown in Figure 2.16 for pure water for which the equation of the straight line is

$$\sigma_W = 0.0757(T_c - T)^{0.776} \tag{2.26}$$

The increase in surface tension of aqueous solutions at 30°C as a function of concentration is shown in Figure 2.17 for the major constituents of geothermal brines. The data in that figure can be approximated by the equation

$$\sigma - \sigma_W = aw_t + bw_t^2 \tag{2.27}$$

Combining the three curves in Figure 2.17 according to the weight ratio of each of the components in a typical geothermal brine as was done for viscosity results in

$$\Delta\sigma = \sigma - \sigma_W = 0.278w_t + 0.0031w_t^2 \tag{2.28}$$

In the absence of further data on salt solutions at higher temperatures, the increase in surface tension can be estimated by

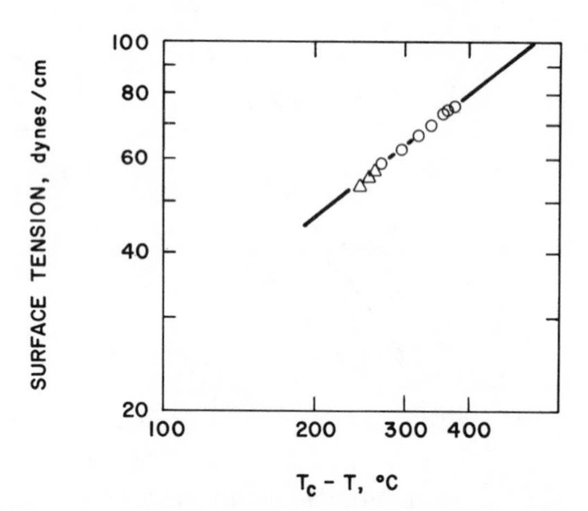

Figure 2.16 Surface tension of water against air as a function of temperature. $T_c = 374.11°C$. Data from *Handbook of Chemistry and Physics* (Weast, 1975, p. F-30) and *International Critical Tables* (Vol. 4, p. 447).

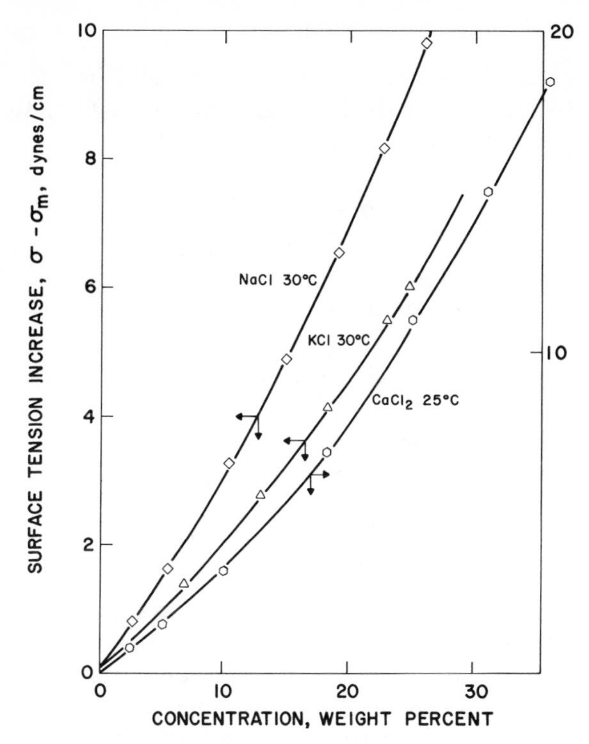

Figure 2.17 Increase in surface tension of water against air due to dissolved salts as a function of concentration of the dissolved salts. Surface tension of water is 71.97 dynes/cm at 25°C. Data from *International Critical Tables* (Vol. 4, p. 465–466).

proportioning the increase in surface tension at 30°C by the ratio of the surface tension of water at the desired temperature to that at 30°C. Doing this and adding the result to the surface tension of water at the desired temperature gives

$$\sigma = \sigma_W \left(1 + \frac{\Delta\sigma_{30}}{\sigma_{W,30}}\right) \tag{2.29a}$$

where the subscript 30 refers to the surface tension data at 30°C. Inserting the surface tension of water as given by Equation 2.26 and at 30°C and the surface tension correction at 30°C as given by Equation 2.28 results in

$$\sigma = 0.00757(T_c - T)^{0.776}(1 + 0.0039w_t + 4.35 \times 10^{-5}w_t^2) \tag{2.29b}$$

Enthalpy

The enthalpy of the brine can be calculated from the foregoing information by integrating the heat capacity over a given temperature range letting enthalpy be 0 at temperature T_0:

$$h = \int_{T_0}^{T} c_B dT \tag{2.30}$$

Using Equation 2.12 for the heat capacity of a brine of constant weight composition and integrating according to Equation 2.30,

$$h = \int_{T_0}^{T} \left[c_W \left(1 - \frac{w_t}{100}\right) + (0.002 + b)w_t \right] dT \tag{2.31}$$

Since the first term of the integral in Equation 2.31 is the enthalpy of pure water corrected for the amount of salts present,

$$h = h_W \left(1 - \frac{w_t}{100}\right) + (0.002 + b)w_t(T - T_0) \tag{2.32}$$

This calculation has been carried out as an example for a 25% total dissolved solids geothermal brine assuming values of b as given by Equation 2.15, and the results are shown in Figure 2.18 together with that of pure water. Multiplying the enthalpy by Equation 2.7 for the density of geothermal brines results in an energy content of the brine per unit volume as a function of temperature as shown in Figure 2.18.

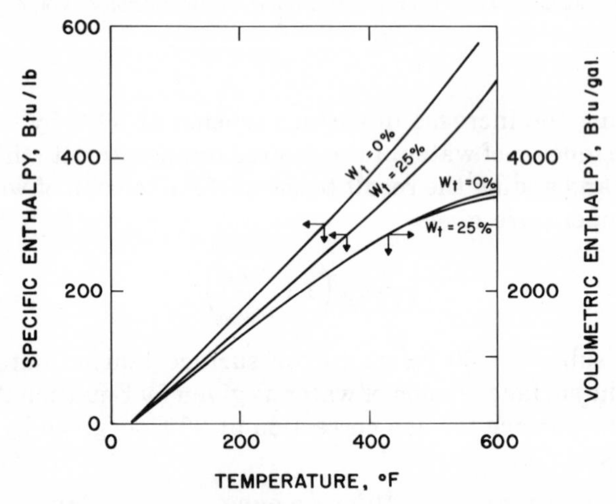

Figure 2.18 Specific enthalpy and volumetric enthalpy of water and a 25% dissolved solids geothermal brine.

This figure shows the effect of decreased density of hot water and the decreased specific heat of salt solutions on the enthalpy of brine. Since wells may be volume flow limited rather than mass flow limited, the gain in power output from a well with increasing temperature may not be as great as thought from a consideration of temperature only because of the decreased density of water. Salt content has little effect on volumetric enthalpy compared with pure water because the increase in density offsets the decrease in heat capacity accompanying increased salt content.

EFFECT OF FLASHING ON BRINE COMPOSITION

The production of geothermal brines by flow of the hydrothermal fluid from the reservoir up through a production well is frequently accompanied by flashing some of the brine due to pressure drop in the upper portions of the well. In general, higher flow rates and higher pressure drops go together so that production wells with high flow rates have a greater degree of flashing of the brine to steam. In some utilization processes, the brine is flashed at the surface to produce steam for use in subsequent processes. This flashing of steam affects the composition of the brine in two ways.

One effect is the simple concentration of the brine as a result of the removal of water by flashing. The greatest concentration change will occur if the brine is flashed to a sufficiently low pressure so that the temperature drops to ambient temperature. The ratio of the concentration of the brine after flashing, w_F, to that before flashing, w_O, is given by the ratio of the water before flashing to that after flashing:

$$\frac{w_F}{w_O} = \frac{1}{(1-y)} \tag{2.33}$$

where y is the weight fraction of water flashed. The fractional change in weight composition, $\Delta w/w_O$, is derived from Equation 2.33:

$$\frac{\Delta w}{w_O} = -y(1-y) \tag{2.34}$$

As an example, the fractional change in weight composition has been calculated for the case of flashing a brine with enthalpy as given by Equation 2.32 to a brine temperature of 150°F. This is shown in Figure 2.19 as a function of temperature. This figure shows that the concentration of solids will increase about 20% for brine flashing from 350 to 150°F.

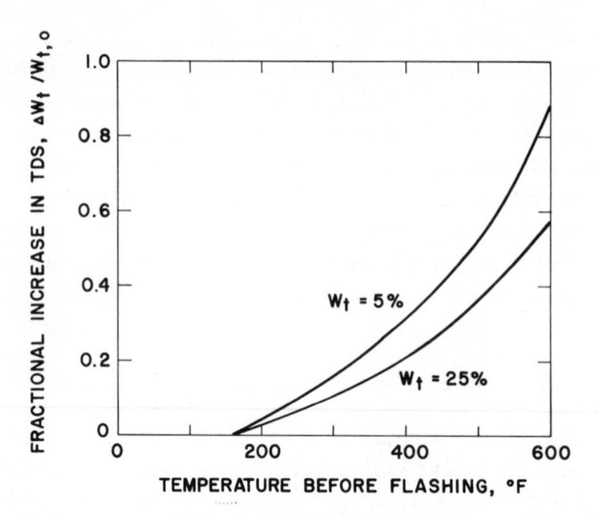

Figure 2.19 Fractional increase in total dissolved solids due to flashing of steam from geothermal brine at the temperature shown to a brine temperature of 150°F.

The other effect of flashing on the chemistry of the brines is the chemical change resulting from the removal of certain constituents such as carbon dioxide or hydrogen sulfide. As carbon dioxide is the dominate gas, this will have the most significant effect on the chemistry. The principle effect of CO_2 release is the pH change. The next most significant effect is the reduction in the dissolved CO_2 in the brine. As far as the calcium carbonate precipitation reaction is concerned, the pH effect is more significant than the change in carbonate concentration itself. This is discussed in Chapter 3.

In addition, the indirect effect of flashing on the equilibria as a result of temperature change will be important in the chemistry of processes for utilizing the geothermal brine.

CHEMICAL PROPERTIES, EQUILIBRIA, AND KINETICS

As described previously, geothermal brines are aqueous solutions of inorganic salts. These salts ionize to form an ionic solution except for silica, which may form a small percentage of nonionic material. The salts in the solution are nearly 100% ionized; consequently, the solution is highly ionic and behaves as such. In addition, some quantity of dissolved gas is present in the brine. This has been discussed previously and so is not considered further in this section. The ionic solution must contain equal cationic and anionic charges. The sum of the cationic and anionic charges have been computed for the typical brines given in Table 2.3 and the results are tabulated in Table 2.4, assuming the ions are present in the ionic form listed in that table. The difference between the two may be due to errors in the chemical analyses, to incomplete analyses leaving out some ions that are present, or to an incorrect assumption as to the ionic form of the elements. However, the discrepancy is not particularly large so that the analysis and the assumptions can be considered representative.

Since the brine starts up the well from the reservoir in equilibrium with solid material, it will be a solution free of precipitates except for entrained solids. As it rises through the well and enters the process equipment, it may contain precipitated salts that have formed during transit from the reservoir to the wellhead. However, since the brine is no longer in contact with the reservoir, the species that precipitate out will not necessarily be those that were present in the reservoir. In addition since the temperature and pressure are not the same as down-hole, the species that limit the solubility will not necessarily be the same. In addition, there may be a pH change that occurs as the material rises up the well further influencing the limiting solubility specie.

Equilibria

The only reactions that occur in a geothermal brine salt solution are simple ionic reactions between cations and anions. An exception to this is the reaction of silica to form higher polymers of silica, which is discussed later. Even so, this is a relatively simple reaction. In general, the equilibria that are involved are given by the reaction of a salt crystal represented by the symbol MX with water to form a solvated cation and anion;

$$MX(s) + aq. = M^{+}aq. + X^{-}aq. \qquad (2.35)$$

In general, the solvated cations and anions must be distributed uniformly throughout the solution to maintain a neutral charge distribution.

Free energy change ΔF is the reversible isothermal work that results from a substance going from state 1 to a state 2. For a perfect gas, this relationship is

$$\Delta F = v dp = RT \ln \left(\frac{P_2}{P_1}\right) \qquad (2.36)$$

For the case of real substances the activity a is defined by

$$\Delta F = RT \ln \left(\frac{a}{a_0}\right) \tag{2.37}$$

For convenience, the activity a_0 in the standard state is defined equal to unity. For the general case of reactions such as

$$aA + bB \rightarrow cC + dD \tag{2.38}$$

the free-energy change is given by

$$\Delta F^\circ = -RT \ln K \tag{2.39}$$

where

$$K = \frac{a_C^c a_D^d}{a_A^a a_B^b} \tag{2.40}$$

This last equation defines the equilibrium constant K in terms of the activities. Thus the equilibrium constant for the equilibria between a solid salt, water, and the aqueous ionic solution as defined by Equation 2.35 is given by

$$K = \frac{a_m a_x}{a_{mx} a_{aq}} \tag{2.41}$$

The relationship of the standard free energy of formation for a chemical reaction to the equilibria constant is shown in Figure 2.20. In that figure a reactant A in its standard state is changed to a product B in its standard state by two paths. One path is the direct one that involves the free energy change ΔF°. The other path involves the change of the reactant A from its standard state to another state, the reversible reaction to the product B at a state of activity other than its standard state, and then the change of B to its standard state. The total work for this latter path is the sum of the three individual terms. The reversible reaction that occurs at equilibria produces zero work and zero free-energy change. Thus, as shown in that figure, the standard free energy change is related to the equilibrium constant for the reaction of A to B at any temperature.

Since the activity of a solid is unity and the activity of water is unity, the equilibrium constant for the Reaction 2.35 is given by

$$K = a_m a_x \tag{2.42}$$

The activity coefficients for any ionic solution are defined by:

$$a = m\gamma \tag{2.43a}$$

$$a = cf \tag{2.43b}$$

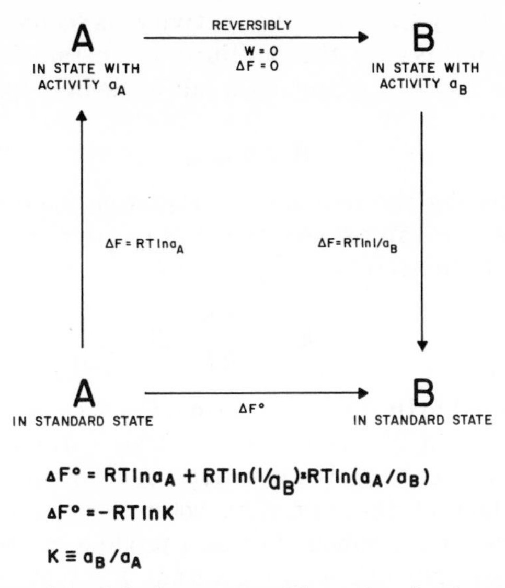

Figure 2.20 Relationship of the free energy of formation of a chemical reaction to the activities of the products and reactants of a chemical reaction at any temperature and the definition of the equilibrium constant K for that reaction.

so that the equilibrium constant is given by

$$K = m_m m_x \gamma_{\pm}^{2} \tag{2.44}$$

where $\gamma_{\pm}$ is the mean molal activity coefficient.

The activity coefficient for the major constituents of geothermal brines is given in Table 2.9. For an ideal solution, the activity

Table 2.9 Mean Molal Activity Coefficient for Aqueous Salt Solutions at 25°C for Salts Which are Major Constituents of Geothermal Brines

Molality	NaCl	KCl	$CaCl_2$
0.01	0.903	0.902	0.732
0.10	0.778	0.770	0.524
1.0	0.656	0.607	0.725
2.0	0.670	0.577	1.554
3.0	0.719	0.572	3.38

Source. Glasstone, 1960.

coefficient is unity so that the activity is equal to the molal concentration. In this case, the equilibrium constant for the solubility or precipitation reaction of the ionic salt as defined by Equation 2.42 becomes

$$K = m_m m_x \tag{2.45}$$

By differentiating the free energy, Equation 2.39, with respect to temperature, it is readily shown that the relationship of the equilibrium constant to temperature is given by

$$\ln K = -\frac{\Delta H}{RT} + B \tag{2.46}$$

where ΔH is equal to the heat of reaction. Consequently, the log of the equilibrium constant versus $1/T$ will plot as a straight line.

For a solubility reaction, the equilibrium constant is known as the solubility product of the salt. The solubility expressed either as solubility of the ionic compound or as a product of the separate ions will plot as a straight line on a log versus $1/T$ plot. As an example, refer to the discussion of the solubility and ionization of silica given in Chapter 4 and of calcium carbonate in Chapter 3.

Kinetics

The rate of ionic reactions is instantaneous as far as the reaction itself is concerned because there is little or no energy barrier that must be overcome to form a precipitate once the ions are in proximity to one another. The reaction of solid salts to form ions in solution is limited by the rate at which the ions from the crystal can diffuse into the solution. The reverse reaction is limited by the rate at which the ions collide to form a crystal in the solution. The mechanism of the formation of a precipitate and the reaction and deposition rate is discussed in Chapter 3. Thus all of the reactions that take place in the geothermal brine will occur essentially instantaneously providing that there is sufficient concentration of the material required for the reaction. An exception to this is the reaction of silica with itself to form higher polymers of silicic acid. This reaction can be a very slow one taking days or months to occur depending on the temperature and pH conditions. Further details are given in the chapter on silica chemistry.

Reactions

In general, the cations whose concentration are determined by equilibria with the alumino-silicates in the reservoir will be undersaturated in the brines at the wellhead because the chlorides of those cations have a very high solubility. Of the anions, chloride is relatively inactive because salts of chloride in general have very high solubilities. Carbonate and the nonionic constituent silica both are relatively reactive species. Carbonate is reactive because of the equilibria between carbon dioxide and hydrogen ion and the reaction of calcium with carbonate to form an insoluble compound. Silica polymerizes to form higher polymers of silicic acid as well as reacting with certain cations to form insoluble salts. The chemistry of both of these constituents is relatively complex, and each is discussed in the following chapters.

Of the cationic constituents, sodium and potassium are present in the most abundant quantities. Both of these constituents have relatively high solubilities with chloride ion and do not react with carbonate, silica, or the other anions that may be present to form insoluble salts or complexes. Thus reactions with these constituents are not important in processes utilizing the brines except for mineral recovery which is discussed later. As mentioned previously, calcium reacts with carbonate and consequently is of importance in the processing of geothermal brines because it may form insoluble deposits that clog pipe and equipment. Its chemistry is discussed in relationship to carbonate chemistry in Chapter 3.

Corrosion

With the exception of silica and its polymers, all of the constituents present in a brine are simple ionic salts, resulting in a solution of electrolytes. Corrosion of equipment processing geothermal brines is generally associated with the hot electrolytic water solution, the presence of hydrogen sulfide gases, and contamination of the brine with oxygen, Table 2.10. In the case of acidic waters, corrosion mechanisms of hydrochloric acid, sulfuric acid, and sulfide become important and are discussed later. However, geothermal brines normally contain no oxygen and thus are in a slightly reduced state. For example, iron may be found in the $+2$ or $+3$ state. As a result, corrosion due to the high sodium chloride content is not often a problem. However, if oxygen does enter the system, then chloride corrosion becomes important.

If the brine is highly acidic or if oxygen enters the system, then high surface corrosion rates will be exhibited on ordinary steels for both the steam and liquid phases. Stress corrosion can also occur. Hydrogen sulfide forms stress corrosion in the form of micro-surface fissures in both steam and liquid phase contact with ordinary steels. Turbine

Table 2.10 Corrosion Characteristics of Various Geothermal Environments

Type of Corrosion	Rate	Phase	Materials Attacked	Components
Surface	1000 mils/yr	Liquid	Steels	Mineral Acids
Surface	100 mils/yr	Steam & Liquid	Steels	Oxygen
Stress	Very High	Steam & Liquid	Steels	Chloride & Oxygen
Stress	Micro-Surface Fissures	Steam & liquid	Steels	Hydrogen Sulfide
Fatigue	--	Steam	Turbine Blades	Hydrogen Sulfide
Surface	Very High	Condensate	Condenser Steel	Hydrogen Sulfide & Oxygen

blades may show fatigue corrosion in the presence of hydrogen sulfide. Surface corrosion will be exhibited on steels in contact with condensate from the steam flashed from geothermal brines because of the presence of hydrogen sulfide and oxygen.

Surface corrosion rates for typical geothermal fluids and different materials are summarized in Table 2.11. This table shows that, as would be expected, copper and its alloys, as well as nickel and monel, exhibit significant corrosion rates in contact with the various geothermal fluids present in a processing plant. Steels, on the other hand, have satisfactory corrosion rates unless oxygen is allowed to enter the system. This is shown in the aerated steam example in which the

Table 2.11 Surface Corrosion Rates in Geothermal Fluid Processes (mil/yr)

		Steels			Copper & its	Nickel &
Geothermal Fluid	Cast Iron	Carbon	Ferritic	Austenitic	Alloys	Monel
Well Water 200°C	1	0.3-0.4	0-0.1	0.1	5-20	6-10
Wellhead Flashed Water, 125°C	0.4	0.3-0.5	0.1[a]	0	0.3-10	1
Wellhead Flash Stream, 100-200°C	1-3	0.3-6	0.3[a]	0	0.3-4	1-4
Steam, Aerated, 100°C	10	18-20	1[a]	0	10-40	8-10
Condensate, 70°C		3-4	0.1[a]	0.1	0.2-5	0.4-4

Source: Marshall, 1973; Lindal, 1974; Dodd, 1975; Foster, 1962.

[a]Pitting.

corrosion rate for carbon steel is about 20 times greater than for steam and water where no oxygen is present. Cast iron shows satisfactory corrosion resistance as it does in seawater.

Table 2.12 summarizes corrosion that occurs and control methods that are used in a geothermal processing plant. Most geothermal brines are neutral or slightly acidic or basic. When quite acidic brines are found, current practice is to not use that brine but drill a new well. Because processing of geothermal brines usually requires separation of flashed steam and gases from the liquid, there will be high velocity gas flow with entrained liquids that may subject the process equipment to erosion. Erosion corrosion rates at the turbine rotor and casing and nozzle diaphragm at Matsukawa, which has acidic brines containing hydrogen sulfide, was determined to be 20 mil/yr (Uchiyama and Matsuura, 1970). Turbine blades will be subjected to hydrogen sulfide stress fatigue as well as erosion caused by droplets of liquids forming in the expanding steam. Carbon steel is satisfactory as a general structural material to resist surface corrosion. Equipment in which air leakage may occur, such as vacuum condensers, ejectors, and cooling towers, will be subjected to surface corrosion caused by aeration of the brine and/or by reaction of hydrogen sulfide to form sulfuric acid. The control methods used with these pieces of equipment are (1) minimize the leakage of air, (2) use sulfuric acid-resistant materials, or (3) neutralize the sulfuric acid. During standby operation of the process equipment, air must be excluded to prevent surface corrosion. External structures subjected to spray of the brine will also corrode. These structures must be built with corrosion resistant materials or coated with corrosion resistant paint. Packings in pumps and valves must be designed to minimize leakage and be of resistant materials to prevent external corrosion. The hydrogen sulfide in the atmosphere surrounding the plant can cause tarnishing of silver surfaces in electronic equipment, which therefore must be protected from this type of corrosion.

External corrosion may also occur in a well. The water surrounding the well casing may be aerated thermal waters or acidic waters. In either case, surface corrosion will occur. The method used in New Zealand to prevent external corrosion of well casings is to protect them with concrete.

SUMMARY

Typical geothermal brines are ionic salt solutions from less than 0.1 up to 40wt.% total dissolved solids and with dissolved gases which are

Table 2.12 Corrosion Occurrence and Control in Geothermal Plants

Equipment Type	Cause of Corrosion	Type of Corrosion	Control Methods
All	Acidic brines	Surface	Minimize use
Well & wellhead	- -	Surface	Carbon steels
Well & wellhead	High velocity	Erosion	Streamlining
Well casing, external	Aerated &/or acidic waters	Surface	Concrete
Turbine blades	Hydrogen Sulfide	Stress fatigue	Low strength steels
Pipelines	- -	Surface	Carbon steels
Condensers, ejectors & cooling towers	H_2S & O_2	Surface	Minimize O_2 leakage; use H_2SO_4 resistant materials; neutralization.
Structures	Spray	Surface	Corrosion resistant materials
Packings	Air & brine	Surface	Minimize leakage; use resistant materials
Electronics	Hydrogen sulfide	Tarnish	Exclude H_2S; use resistant materials
Standby operation	Air & brine	Surface	Exclude air

about 90% CO_2. The total dissolved solids are 70 to 80wt.% sodium chloride, a sodium to potassium atomic ratio of 10 : 1, less than 8wt.% calcium chloride, and 200 to 800 ppm silica, all other constituents being relatively negligible.

The physical properties of brine can be approximated by the properties of pure water and the weight percent of total dissolved solids as follows:

Density:

$$\rho = \rho_W + 0.0073 w_t [1 + 1.6 \times 10^{-6} (T - 273)^2] \quad (2.7)$$

Heat capacity:

$$c = c_w \left(1 - \frac{w_t}{100}\right) - 0.002 w_t \quad (2.15a)$$

See also Equations 2.14, 2.15b, and 2.15c.

Vapor pressure:

$$p = p_W \left(1 - \frac{0.004 w_t}{\rho}\right) \quad (2.21)$$

Surface tension:

$$\sigma = 0.0757 (T_c - T)^{0.776} (1 + 0.0039 w_t + 4.35 \times 10^{-5} w_t^2) \quad (2.29b)$$

Viscosity:

$$\mu = \mu_W (1 + 0.021 w_t + 0.00027 w_t^2) \quad (2.25)$$

Enthalpy:

$$h = h_W \left(1 - \frac{w_t}{100}\right) - 0.007 w_t (T - T_0) \quad (2.32)$$

Chemical equilibria are related to the standard free energy of formation. The ratio of product and reactant concentrations can be calculated using the equilibrium constant. The variation of the equilibrium constant is conveniently represented by a log concentration versus reciprocal absolute temperature plot, which is typically a straight line with slope proportional to the heat of solution. The activity coefficient is less than one, and the effect of ionic strength on the activity coefficient must be considered.

NOMENCLATURE

a activity
b arbitrary constant
B constant of integration
$\bar{c}$ partial molal heat capacity, Btu/lb mole °F or cal/g-mol °C
c heat capacity, Btu/lb °F, or cal/g °C, subscript designates material
C_i molecular concentration of the ith constituent, mol/liter
F free energy, ΔF° standard free energy of a reaction
f mole concentration activity coefficient
H_s enthalpy change on dissolving salt, cal/g mol
K_{eq} equilibrium, constant
m molality, mol/kg of solution
p vapor pressure of water over brine, atm
P total pressure, atm
P_i vapor pressure of the pure component i, atm
R gas constant
T temperature, °K
w_i weight concentration of ith constituent parts per hundred, that is, wt.%
w_t weight concentration of total dissolved solids, parts per hundred, that is, wt.%
x mole fraction of component i
y weight fraction
z valence of the ions in solution

Greek Symbols

α activity coefficient
γ molal activity coefficient
μ viscosity, centipoise
ρ density, g/cm^3
σ surface tension, dyn/cm

Subscripts

B brine
i ith component of the solute; i^+ for cations, i^- for anions

s solute, single component
t total dissolved solids
T temperature, *T*
W water, pure

REFERENCES

Composition

Arnorsson, S., "Underground Temperatures in Hydrothermal Areas in Iceland as Deduced from the Silica Content of the Thermal Water," U.N. Symposium on Development and Utilization of Geothermal Resources, *Geothermics*, Special Issue 2, **2**, 536 (1970).

Arnorsson, S., "The Distribution of Some Trace Elements in Thermal Waters in Iceland," U.N. Symposium on Development and Utilization of Geothermal Resources, *Geothermics*, Special Issue 2, **2**, 542 (1970*a*).

Arnorsson, S., "Geochemical Studies on Thermal Waters in the Southern Lowlands of Iceland," U.N. Symposium on Development and Utilization of Geothermal Resources, *Geothermics*, Special Issue 2, **2**, 547 (1970*b*).

Arnorsson, B., and J. Tomasson, "Deuterium and Chloride in Geothermal Studies in Iceland," U.N. Symposium on Development and Utilization of Geothermal Resources, *Geothermics*, Special Issue 2, **2**, 1405 (1970).

Bjornsson, S., S. Arnorsson, and J. Tomasson, "Exploration of the Reykjanes Thermal Brine Area," U.N. Symposium on Development and Utilization of Geothermal Resources, *Geothermics*, Special Issue 2, **2**, 1640 (1970).

Dal Secco and G. Alfredo, "Geothermal Plants Gas Removal from Jet Condensers," Second U.N. Symposium on Development and Utilization of Geothermal Resources, San Francisco, May 1975.

Dodd, F. J., "Material and Corrosion Testing at the Geysers Geothermal Power Plant," Second U.N. Symposium on Development and Utilization of Geothermal Resources, San Francisco, May 1975.

Dominco, E., and E. Samilgil, "The Geochemistry of the Kizildere Geothermal Field, in the Framework of the Saraykoy–Denizli Geothermal Area," U.N. Symposium on Development and Utilization of Geothermal Resources, *Geothermics*, Special Issue 2, **2**, 553 (1970).

Ellis, A. J., "Volcanic Hydrothermal Areas and the Interpretation of Thermal Water Compositions," Iav. Symposium, New Zealand, 1965.

Ellis, A. J., "Present Day Hydrothermal Systems and Mineral Deposition," Proc. 9th Commonwealth Mining Metallurgical Congress, London, 1969.

Ellis, A. J., "Quantitative Interpretation of Chemical Characteristics of Hydrothermal Systems," U.N. Symposium on Development and Utilization of Geothermal Resources, *Geothermics*, Special Issue 2, **2**, 516 (1970).

Fujii, Y., and T. Akeno, "Chemical Prospecting of Steam and Hot Water in the

Matsukawa Geothermal Area," U.N. Symposium on Development and Utilization of Geothermal Resources, *Geothermics*, Special Issue 2, **2**, 1416 (1970).

Glover, R. B., "The Chemistry of Thermal Waters at Rotorua," *N. Z. J. Sci.*, **10**, 70–96 (1967).

Hayashida, T., and Y. Ezima, "Development of Otake Geothermal Field," U.N. Symposium on Development and Utilization of Geothermal Resources, *Geothermics*, Special Issue 2, **2**, 208 (1970).

Helgeson, H. C., "A Chemical and Thermodynamic Model of Ore Deposition in Hydrothermal Systems," 50th Anniversary Symposia, Mineralogical Society of America, Special Publication No. 3, 1970.

Koga, A., "Geochemistry of the Waters Discharged from Drillholes in the Otake and Hatchobaru Areas," U.N. Symposium on Development and Utilization of Geothermal Resources, *Geothermics*, Special Issue 2, **2**, 1422 (1970).

Mahon, W. A. J., and Klyen, "Chemistry and the Tokaanu-Waihi Hydrothermal Area," *N. Zealand J. Sci.*, **11**, 140–158 (1968).

Mahon, W. A. J., and J. B. Finlayson, "The Chemistry of The Broadlands Geothermal Area New Zealand," *Amer. J. Sci.*, **272**, 48–68 (January 1972).

Mercado, S., "High Activity Hydrothermal Zones Detected by Na/K, Cerro Prieto, Mexico," U.N. Symposium on Development and Utilization of Geothermal Resources, *Geothermics*, Special Issue 2, **2**, 1367 (1970).

Nakamura, H., K. Sumi, K. Katagiri, and T. Iwata, "The Geological Environment of Matsukawa Geothermal Area, Japan," U.N. Symposium on Development and Utilization of Geothermal Resources, *Geothermics*, Special Issue 2, **2**, 221 (1970).

Ragnars, K., K. Saemundsson, S. Benediktsson, and S. S. Einarsson, "Development of the Namafjall Area, Northern Iceland," U.N. Symposium on Development and Utilization of Geothermal Resources, *Geothermics*, Special Issue 2, **2**, 925 (1970).

Sato, K., "The Present State of Geothermal Development in Japan," U.N. Symposium on Development and Utilization of Geothermal Resources, *Geothermics*, Special Issue 2, **2**, 155 (1970).

Skinner, B. J., et al., "Sulfides Associated with Salton Sea Geothermal Brine," *Econ. Geol.*, **62**, 316 (1967).

Swanberg, C. A., "The Application of the Na–K–Ca Geothermometer to Thermal Areas of Utah and the Imperial Valley, California," *Geothermics*, **3**, 53 (1974).

Sigvaldason, G. E., and G. Cuellar, "Geochemistry of the Ahuachapan Thermal Area, El Salvador, Central America," U.N. Symposium on Development and Utilization of Geothermal Resources, *Geothermics*, Special Issue 2, **2**, 1392 (1970).

Thorhallsson, Sverrir, et al., "Rapid Scaling of Silica in Two District Heating Systems," Second U.N. Symposium on Development and Utilization of Geothermal Resources, San Francisco, May 1975.

Wahl, E. F., I. K. Yen, and W. J. Bartel, "Silicate Scale Control in Geothermal Brines," Contract No. 14-30-3041, Office of Saline Water, U.S. Department of the Interior, Washington, D.C., September 1, 1974.

Corrosion

Dodd, F. J., et al., "Material and Corrosion Testing at the Geysers Geothermal Power Plant," Second U.N. Symposium on Development and Utilization of Geothermal Resources, San Francisco, May 1975.

Foster, P. K., "The Thermodynamic Stability of Iron and Its Compounds in Hydrothermal Media," *N.Z.J. Sci.*, **2**, 422–430 (1959).

Foster, P. K., and A. Tombs, "Corrosion by Hydrothermal Fluids," *N.Z.J. Sci.*, **5**, 28–42 (1962).

Hanck, J. A. and G. Nekoksa, "Corrosion Rate Monitoring at the Geysers Geothermal Power Plant," Second Second U.N. Symposium on the Development and Use of Geothermal Resources, San Francisco, May 1975.

Marshall, T., and W. R. Braithwaite, "Corrosion Control in Geothermal Systems," *Geothermal Energy*, UNESCO, Paris, 1973.

McDowell, G. D., "The Scrubbing of Chlorides in Carry-over Water from Geothermal Well Separators," Second U.N. Symposium of the Development and Use of Geothermal Energy, San Francisco, May 1975.

Toliva, M. E., "Corrosion of Turbine Materials in Geothermal Steam Environments in Cerro Prieto, Mexico," Second U.N. Symposium on the Development and Use of Geothermal Resources, San Francisco, May 1975.

Uchiyama, M., and S. Matsuura, "Measurement and Transmission of Steam in Matsukawa Geothermal Power Plant," *Geothermics*, Special Issue 2, **2** (2), 1572–1580 (1970).

Physical Properties

Dorsey, N. E., *Properties of Ordinary Water Substance*, A.C.S. Monograph Series No. 81, Huffner, New York, 1968.

Ellis, A. J., and R. M. Goldwig, "The Solubility of Carbon Dioxide above 100°C in Water and in Sodium Chloride Solutions," *Amer. J. Sci.*, **261**, 47–60 (1963).

Fabuss, B. M., and A. Korosi, "Thermodynamic Properties of Sea Water and its Concentrates," *Desalination*, **2**, 271–278 (1967).

Gardner, W. L., and R. E. Mitchell, and J. W. Cobble, "Thermodynamic Properties of High Temperature Aqueous Electrolytes," *J. Phys. Chem.*, **73**, 2025 (1969).

Gardner, W., "Calorimetry and Thermodynamics of Aqueous Electrolytes at High Tempearatures," Ph.D. Thesis, Purdue Univ., 1968, University Microfilms No. 69–2921.

Glasstone, S., and D. Lewis, *Elements of Physical Chemistry*, van Nostrand, Princeton, N.J., 1960.

Haas, J. L., "An Equation for the Density of Vapor-Saturated NaCl–H_2O Solutions from 75° to 325°C," *Amer. J. Sci.*, **269**, 489–493 (1970).

Keenan, J. H., and F. G. Keyes, *Thermodynamic Properties of Steam*, Wiley, New York, 1951.

Likke, S., and L. A. Bromley, "Heat Capacities of Aqueous NaCl, KCl, $MgCl_2$, $MgSO_4$, and Na_2SO_4 Solutions between 80°C and 200°C," *J. Chem. Eng. Data*, **18**, 189–195 (1973).

Liu, C., and W. T. Lindsay, "Thermodynamic Properties of Aqueous Solutions at High Temperatures," U.S. Department of the Interior R&D Report No. 722 (December 1971).

Perry, R. H., and C. H. Chilton, *Chemical Engineers Handbook, fifth ed.*, McGraw-Hill, New York, 1973.

Seidell, A., *Solubilities of Inorganic and Metal-Organic Compounds*, 4th ed., Amer. Chem. Soc., New York, 1965.

Silvester, L. F., et al., "Thermodynamics of Geothermal Brines I. Thermodynamic Properties of Vapor-Saturated NaCl(aq) Solutions from 0–300°C," LBL-4456, Lawrence Berkeley Laboratory, Berkeley, Calif., 1976.

Weast, R. C., Ed., *Handbook of Chemistry and Physics*, 56th ed., Chemical Rubber Publishing Co., Cleveland, Ohio, 1975.

CHAPTER 3

Chemistry of Carbonates as Related to Geothermal Brine Utilization

The significant reactions of carbonates in geothermal brines involve the release of carbon dioxide, a change in pH, and the formation of insoluble precipitates. These are readily understood in terms of the reactions that occur in the interesting carbon dioxide–carbonate–bicarbonate water system on which the reactions depend.

THE CARBON DIOXIDE-WATER SYSTEM

Carbon dioxide is a linear molecule, the carbon and two oxygen atoms arranged in a straight line. The bonds between the carbon and oxygen atoms are covalent double bonds. Resonance occurs in the molecule, however, so that there is a tendancy for triple bonds to occur at one end of the molecule with a single bond on the other side, thus giving rise to a polar molecule. The result of the resonance is an above average bond energy and smaller bond distances between the carbon and oxygen atoms. However, the polarity is relatively small so that in interactions with other molecules, it behaves essentially as a nonpolar molecule.

Gas Solubility

If carbon dioxide gas is brought into contact with pure water, there will be a transfer of carbon dioxide molecules from the gas into the liquid forming a solution of dissolved carbon dioxide molecules in the water. The dissolved carbon dioxide, that is aqueous carbon dioxide, is probably present both as CO_2 molecules and as CO_2 molecules associated with one water molecule forming unionized carbonic acid with the formula H_2CO_3. Assuming that the ratio of these two species is constant in solution because the activity coefficients of nonpolar substances in water do not change, then the solubility of carbon dioxide in water will be given by Henry's law. Henry's law states that in dilute solutions the concentration of the dissolved gas will be proportional to its partial pressure in a gas phase. Another way of stating this is that the free energy and therefore activity of the dissolved CO_2 in the liquid must equal the free energy and therefore activity of CO_2 in the gas phase. At normal temperatures and pressures, the activity of a gas is equal to its partial pressure so that the activity in the liquid phase must equal the partial pressure in the gas phase. Since the activity of the carbon dioxide in the liquid phase will be proportional to its concentration and, assuming a constant activity coefficient, the Henry's law relationship that the partial pressure is proportional to the concentration in the liquid phase applies.

The relationship already discussed for the solution of CO_2 gas in water to form aqueous CO_2 or carbonic acid is shown in Table 3.1. In that table, Reaction 3 has been written to indicate the sum of reactions 1 and 2 so that the equilibrium constant represents a measure of the CO_2 dissolved in the liquid both as aqueous CO_2 and carbonic acid. Consequently, the equilibrium constant for that reaction is related to the equilibrium constant for each of the individual reactions by

$$K_H = \frac{[(CO_2aq) + (H_2CO_3)]}{p_{CO_2}} \tag{3.1a}$$

$$K_H = K_x\left(1 + \frac{K_y}{K_x}\right) \tag{3.1b}$$

Note that in principle Henry's law applies only to Reaction 1 but that since the ratio of carbonic acid to CO_2 for Reaction 2 should be a constant, then Henry's law will also apply to the concentration of the dissolved CO_2 where dissolved CO_2 means both aqueous CO_2 and carbonic acid. In this case, Henry's law constant K_H will be the same

Table 3.1 Reactions, Equilibria, and Equilibrium Constants for the Carbonate-Water System

Reaction			Equilibrium Constant	
Type	No.	Reaction	Definition	Value
Gas solubility	1	$CO_2(g) + aq = CO_2\ aq$	$K_x = (CO_2\ aq)/p_{CO_2}$	
	2	$CO_2\ aq + H_2O = H_2CO_3$	$K_y = (H_2CO_3)/(CO_2)$	
	3	$CO_2(g) + H_2O = CO_2 \cdot H_2O$	$K_H = (CO_2)/p_{CO_2}$	1600 at 25°C
Ionization	4	$H_2CO_3 = HCO_3- + H^+$	$K_1 = (H^+)(HCO_3-)/(H_2CO_3)$	4.47×10^{-7} at 25°C
	5	$HCO_3^- = CO_3^{2-} + H^+$	$K_2 = (H^+)(CO_3^{2-})/(HCO_3^-)$	5.62×10^{-11} at 25°C
Hydrolysis	6	$CO_3^{2-} + H_2O = HCO_3^- + OH^-$	$K_6 = K_1K_w$	
	7	$HCO_3^- + H_2O = H_2CO_3 + OH^-$	$K_7 = K_2K_w$	
Acid-base	8	$CO_3^{2-} + H_2CO_3 = 2HCO_3^-$	$K_8 = K_1/K_2$	0.8×10^4, at 25-250°C
HCO_3^- as base	4R	$HCO_3^- + H^+ \rightarrow H_2CO_3$		
HCO_3^- as acid	6R	$HCO_3^- + OH^- \rightarrow H_2O + CO_3^{2-}$		
CO_2 release	9	$HCO_3^- \rightarrow CO_2 + OH^-$		

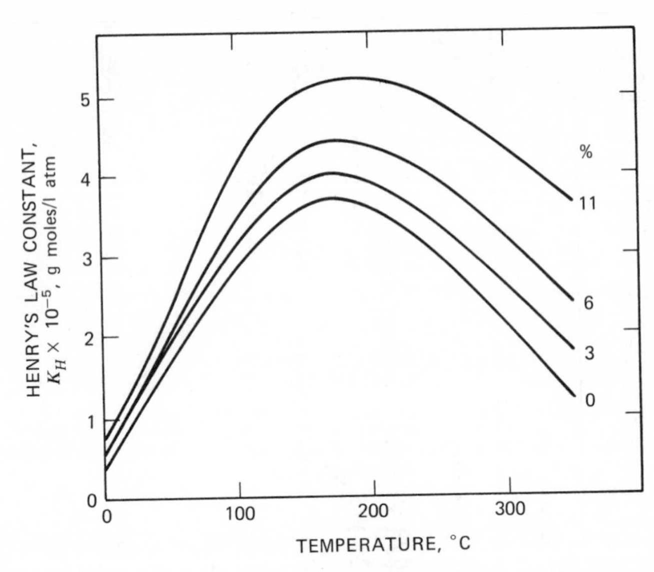

Figure 3.1 Henry's law constant for the solubility of carbon dioxide in water and sodium chloride solutions for various concentrations of sodium chloride as shown on the curves in weight percent. From Ellis (1963*a*). The constant is independent of pressure up to 50 atm.

as the constant given in Equation 3.1*b*. Its value as a function of temperature is shown in Figure 3.1 for pure water and salt concentrations up to 11wt.%. Since geothermal brines are mainly sodium chloride, Table 2.1, this figure can be used to estimate the solubility of carbon dioxide in brine. Figure 3.1 is valid for temperatures up to 650°F and pressures up to 50 atm, and so should be sufficient for most design purposes.

Ionic Reactions

The ionization, hydration, and acid–base reactions that take place in the carbon dioxide water system are summarized in Table 3.1.

Carbonic acid dissociates, that is ionizes, to form bicarbonate ion and hydrogen ion as shown by Reaction 4 in Table 3.1. The equilibrium constant K_1 for this reaction is called the first ionization constant of carbonic acid. The bicarbonate ion, in turn, will ionize as shown by Reaction 5 to form carbonate ion and hydrogen ion. The equilibrium constant for this reaction is the second ionization constant of carbonic acid as shown in Table 3.1.

Both carbonate and bicarbonate ions will undergo hydration to release a hydroxide ion as shown by Reactions 6 and 7 in Table 3.1. The equilibrium constants for these reactions are related to the disociation constant K_W for water and the appropriate ionization constant of carbonic acid from Reactions 4 and 5 in Table 3.1. Carbonic acid is strongly hydrolized so that if there is any appreciable quantity of carbonate in solution then the solution will be basic.

The bicarbonate ion can act either as a base or as an acid. This is shown by writing Reactions 4 and 6 in the reverse direction as shown in Table 3.1. As a base, bicarbonate reacts according to Equation 4R and as an acid according to Reaction 6R. Reaction 8 in that table shows the acid–base reaction that occurs between carbonate and bicarbonate ion. As that reaction is written, carbonate is acting as a base because it picks up hydrogen ions, and carbonic acid is acting as an acid because it is releasing a hydrogen ion to form bicarbonate.

The significance and interaction of these reactions is demonstrated by the pH change of a carbonate, bicarbonate, or carbonic acid solution as either a strong acid or strong base is added to it. Referring to Figure 3.2, the acid that is added to a basic solution of carbonate will be consumed by the carbonate until it is almost completely converted to bicarbonate. At that point, the pH will change rapidly with a small addition of acid until a pH of about 7 is reached. Then the pH will remain relatively constant in the neighborhood of 5 to 6 until all of the bicarbonate is consumed. At this point, the pH will change

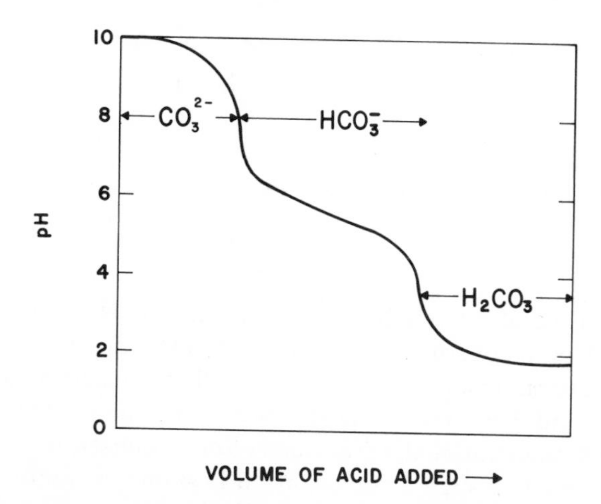

Figure 3.2 The change in pH of a sodium carbonate solution with the addition of acid. The addition of base would give the same curve but reading from right to left, the opposite direction of the acid addition.

rapidly with the addition of a small amount of acid to a pH of about 2. The addition of base would behave in the reciprocal fashion. This example shows that in a bicarbonate solution with a pH of about 7 the bicarbonate acts as a buffer to the addition of acid or base by consuming the acid or base according to the Reaction 4R or 6R in Table 3.1 so as to maintain the pH near 7. If a geothermal brine containing bicarbonate with a pH in the neighborhood of 7 flashes, the initial CO_2 release, which will proceed according to Equation 9 in Table 3.1, will form hydroxide ions so tending to make the solution more basic. However, the hydroxide ion will be buffered by the presence of the remaining bicarbonate. Thus the change in pH accompanying the initial CO_2 release will not be as large as would be the case with more complete flashing and release of large quantities of carbon dioxide. The extent to which the solution becomes basic upon flashing will depend on the initial concentration of bicarbonate and therefore the amount of hydroxide ion that is produced by release of carbon dioxide. For example, a neutral bicarbonate solution containing 100 ppm of bicarbonate would be capable of releasing 1.6 mmol of hydroxide ion/liter.

The analyses of geothermal brines generally report the total carbon dioxide concentration as a single number that represents the sum of the dissolved carbon dioxide plus the bicarbonate plus the carbonate. This may be expressed as bicarbonate or carbonate ion. If the total carbon dioxide present in the water solution is reported as bicarbonate and is defined by

$$C_T = (CO_2 \cdot H_2O) + (HCO_3^-) + (CO_3^{2-}) \tag{3.2}$$

then the weight percent of carbon dioxide contained in a water system reported as bicarbonate will be given by

$$w_{HCO_3} = 0.1C_T(MW_{HCO_3}) \tag{3.3}$$

Another method of visualizing the interaction of the reactions shown in Table 3.1 is to compute the fraction of carbonic acid, bicarbonate, and carbonate that is present at any pH. This can be done using the first and second dissociation constants of carbonic acid together with the fact that the sum of the fractions of each constituent must equal unity. The result of such a computation is shown in Figure 3.3. Since the pH of geothermal brines is typically in the range of 6 to 8, it is evident from Figure 3.3 that most of the carbon dioxide species are present as bicarbonate.

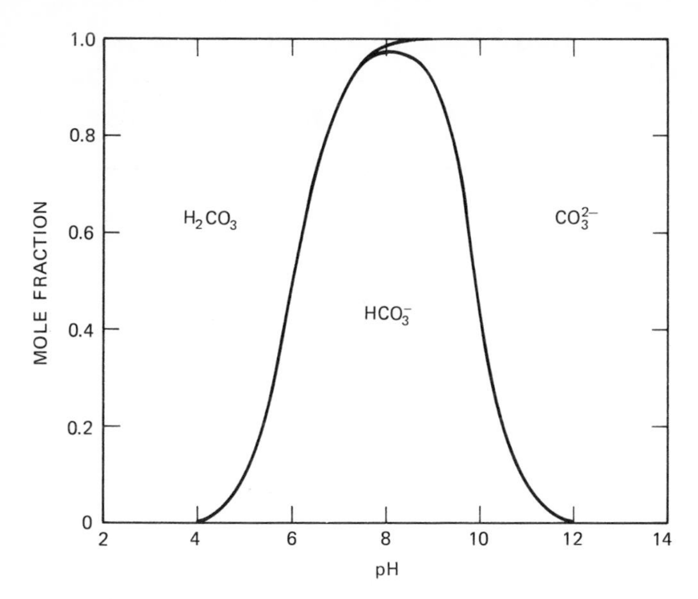

Figure 3.3 Fraction of carbonate, bicarbonate, carbonic acid, and CO_2 present as a function of pH at 100°C.

CATIONIC REACTIONS OF BICARBONATE SOLUTIONS

The carbonates and bicarbonates of the alkali earth metals are highly soluble as are the other anionic compounds of this group. The bicarbonates of the other cations found in geothermal brines such as barium, calcium, lead, magnesium, strontium, and iron are essentially insoluble as indicated by the solubility products given in Table 3.2. It will be noted from this table that magnesium carbonate has a higher solubility than calcium carbonate, and furthermore since magnesium is generally present in smaller quantities than calcium in geothermal brines, calcium carbonate will precipitate out rather than magnesium. Although their carbonates are less soluble, iron, lead, manganese, strontium, and barium are all present in substantially less quantities than calcium; thus calcium carbonate is the material that usually precipitates.

CHEMISTRY OF CALCIUM CARBONATE AS RELATED TO DEPOSITION

Solid calcium carbonate occurs in two crystalline structures: aragonite and calcite. Divalent cations with a large ionic radius such as

Table 3.2 Solubility Product of Carbonates

Carbonate	Solubility Product	Temperature (°C)
Barium	7×10^{-9}	16
Calcium	1.0×10^{-8}	15
Ferrous	2×10^{-11}	--
Lead	3.3×10^{-14}	18
Magnesium	2.6×10^{-5}	12
Manganous	9×10^{-11}	--
Strontium	1.6×10^{-9}	25

barium and lead crystallize in the aragonite form. The calcite–aragonite phase transition diagram, Figure 3.4, shows that calcite is the equilibrium form to be expected at process conditions in geothermal utilization plants. However, either or both can be precipitated at normal conditions from aqueous solutions depending on the

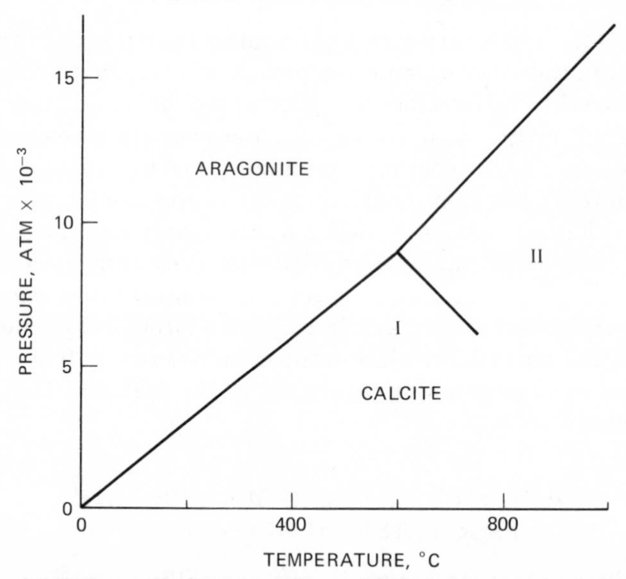

Figure 3.4 Argonite–calcite phase transition diagram. Two calcite crystal structures are shown on the diagram as indicated by I and II. From Boettcher and Wyllie (1967).

temperature, pressure, reactant concentrations, and other ionic species present. Carbonate concentration, lead chloride, magnesium nitrate or chloride, and cobalt nitrate have been shown to effect the ratio of aragonite to calcite produced under laboratory conditions (Kirov and Filizova, 1970; Doner and Pratt, 1969). The ratio also depends on the degree of supersaturation (Girou and Rogues, 1969).

Equilibrium Considerations

The principle chemical reactions describing calcium carbonate deposition are:

1. The precipitation of calcium carbonate:

$$Ca^{2+} + CO_3^{2-} = CaCO_3 \tag{3.4}$$

for which the solubility product is

$$K_c = (Ca^{2+})(CO_3^{2-}) \tag{3.5}$$

2. The equilibrium reactions between the carbonate, bicarbonate, hydrogen ions, and carbon dioxide:

$$CO_2{\cdot}H_2O = HCO_3^- + H^+ \tag{3.6}$$

$$HCO_3^- = H^+ + CO_3^{2-} \tag{3.7}$$

for which the solubility products are, respectively,

$$K_2 = \frac{(H^+)(CO_3^{2-})}{(HCO_3^-)} \tag{3.8}$$

$$K_1 = \frac{(H^+)(HCO_3^-)}{\alpha_{CO_2}} \tag{3.9}$$

where α_{CO_2} is the activity of carbon dioxide in the solution.

Combining the foregoing yields the overall equilibrium

$$Ca^{2+} + CO_2 + H_2O = CaCO_3 + 2H^+ \tag{3.10}$$

with the overall solubility product

$$K_c = \frac{(Ca^{2+})(\alpha_{CO_2})K_1K_2}{(H^+)^2} \tag{3.11}$$

Use can be made of the Henry's law relationship, Equation 3.1*b*, and the activity coefficient, Equation 2.43, to determine the activity of carbon dioxide in the solution:

$$\alpha_{CO_2} = \frac{\gamma P_{CO_2}}{k} \tag{3.12}$$

Equation 3.11 shows that the calcium carbonate solubility product can be determined experimentally from three concentrations, such as the partial pressure of CO_2, the calcium concentration, and the hydrogen ion concentration. If the experimental determination is carried out with pure water and calcium bicarbonate so that the bicarbonate concentration is twice the calcium concentration, then by appropriate combination of the equilibrium relationships

$$K_c = \frac{4(Ca^{2+})^3 \gamma_\pm^{\ 3} K_2}{K_1 \alpha_{CO_2}} \tag{3.13}$$

This assumes that the carbonate and CO_2 concentrations are small compared to bicarbonate and can be neglected in material balance calculations. From this equation and data for the calcium or bicarbonate concentration and the activity of CO_2, the solubility product can be determined. The experimentally determined solubility product is given in Figure 3.5 and Table 3.3.

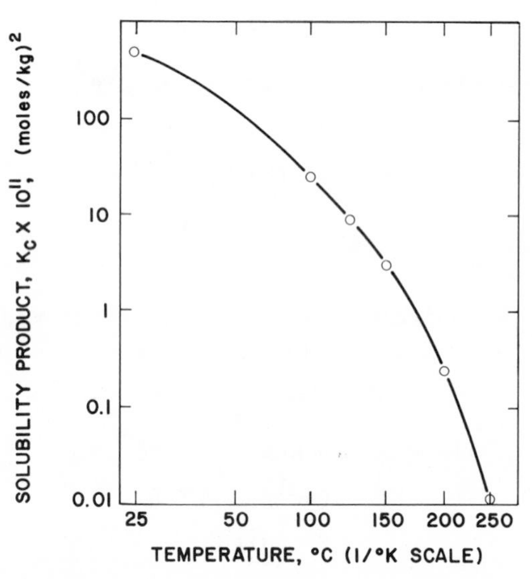

Figure 3.5 The solubility product of calcite as a function of temperature. Data from Ellis (1963*b*).

Table 3.3 Solubility Product of Calcium Carbonate from Various Sources

Molal Solubility Product	Temperature	Source
4.8×10^{-9}	25°C	Ellis, 1963
$2.8\text{–}4.0 \times 10^{-9}$	25°C	Nakayama, 1971

The concentration of calcium in equilibrium with calcite for a given bicarbonate concentration is obtained by combining the equilibrium relationships, Equations 3.5, 3.8, and 3.9, to give

$$(\mathrm{Ca}^{2+}) = \frac{K_c \alpha_{\mathrm{CO}_2} K_1}{K_2(\mathrm{HCO}_3^-)^2} \tag{3.14}$$

In terms of the pH of the solution and the carbon dioxide concentration, the relationship is

$$(\mathrm{Ca}^{2+}) = \frac{K_c(\mathrm{H}^+)^2}{(\alpha_{\mathrm{CO}_2})K_1K_2} \tag{3.15}$$

Making use of Equation 3.6, the calcium concentration in terms of bicarbonate and pH is

$$(\mathrm{Ca}^{2+}) = \frac{K_c(\mathrm{H}^+)}{(\mathrm{HCO}_3^-)K_2} \tag{3.16a}$$

The bicarbonate concentration is related to pH and the total CO_2 present as dissolved CO_2, carbonate, or bicarbonate ion by Equations 3.2, 3.8, and 3.9:

$$C_T = (\mathrm{HCO}_3^-)\left[1 + \frac{(\mathrm{H}^+)}{K_1}\right] + \frac{(\mathrm{H}^+)K_2}{(\mathrm{HCO}_3^-)} \tag{3.17}$$

Thus the calcium ion concentration can also be expressed by

$$(\mathrm{Ca}^{2+}) = \frac{K_c(\mathrm{H}^+)}{f_1[(\mathrm{H}^+)(C_T)]K_2} \tag{3.16b}$$

where f_1 is the solution of Equation 3.17 for (HCO_3^-).

Effect of Other Ionic Constituents

Magnesium ion as well as other factors have been shown to affect the precipitation and dissolution rates of calcite (Weyl, 1967; Berner, 1967; Biscoff, 1968; Doner and Pratt, 1969). For geothermal brines, the effect of ionic strength on the solubility product can be estimated from Figure 3.6, and the relationship

$$K_{cc} = K_c \gamma_{\pm}^{3} \tag{3.18}$$

This figure shows that the activity coefficient increases below 0.6 molar, that is, 3wt.%, and, therefore, the solubility of calcite depends markedly on the NaCl concentration. Varying the total solids concentration of seawater from 3 to 10wt.% does not change the calcium carbonate solubility product (Langelier, et al. 1950). To use Figure 3.6, the total ionic strength μ is calculated by

$$\mu = \frac{1}{2} \sum_i m_i z_i^2 \tag{3.19}$$

The corrected solubility product is then used in Equations 3.14 to 3.16*b*. Also, note that in the foregoing equations, molality may be used rather than mole concentration and is frequently more convenient. For more concentrated solutions, the use of molality with molal activity coefficient is the correct procedure as described by Equations 2.42 to 2.44.

Although not an ionic constituent, silica may increase the solubility of calcite due to the formation of complexes between calcium ion and silica.

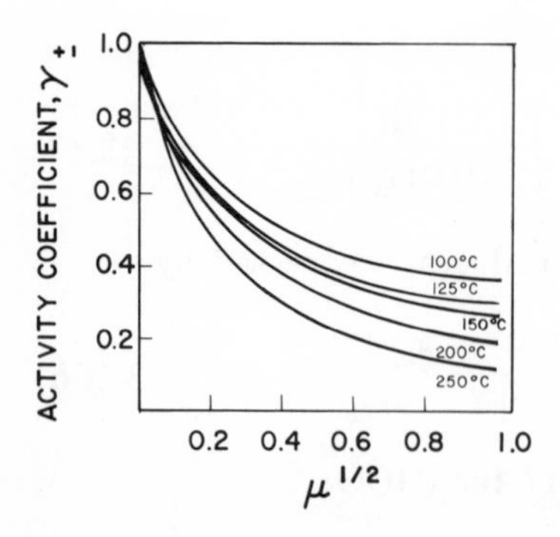

Figure 3.6 Mean molal activity coefficient for calcium bicarbonate as a function of total ionic strength for various temperatures. From Ellis (1963*b*).

Effect of Temperature

There is considerable data in the literature on the solubility of $CaCO_3$ at ordinary temperatures and pressures. The solubility of calcite in water from 100 to 275°C is given in Table 3.4. The equation relating the equilibrium constant to temperature,

$$\ln K_c = -\left(\frac{\Delta H}{R}\right)\left(\frac{1}{T}\right) \tag{3.20}$$

can be used to determine the heat of solution for calcite using the data of Figure 3.5. As deduced from the figure, ΔH is not constant but varies from 11 to 31 kcal/g-mol.

Table 3.4 Solubility of Calcite in Water-Carbon Dioxide Solutions at Given Values of Temperature and CO_2 Pressure; Solubility in mmol $CaCO_3$/kg H_2O

p_{CO_2}	Temperature (°C)							
(atm)	100	125	150	175	200	225	250	275
1	2.16	1.42	0.94	0.60	0.40	0.27	0.15	0.08
4	3.60	2.44	1.58	0.97	0.63	0.39	0.24	0.13
12	5.55	3.57	2.21	1.44	0.91	0.59	0.36	0.20
62	----	----	4.05	2.55	1.52	0.89	0.51	0.28

Source. From Ellis, 1963b.

Effect of Pressure

Experimental data for the solubility product K_c of $CaCO_3$ in water between 1 and 25°C is shown in Figure 3.7 as a function of pressure. The change in the solubility product with pressure as derived from equilibrium thermodynamic conditions is given by

$$\frac{\delta(\ln K_c)}{\delta P} = -\frac{\Delta V}{RT} \tag{3.21}$$

where ΔV is the difference in product-reactant molal volumes. Using the data shown in Figure 3.7 and Equation 3.21, experimental values

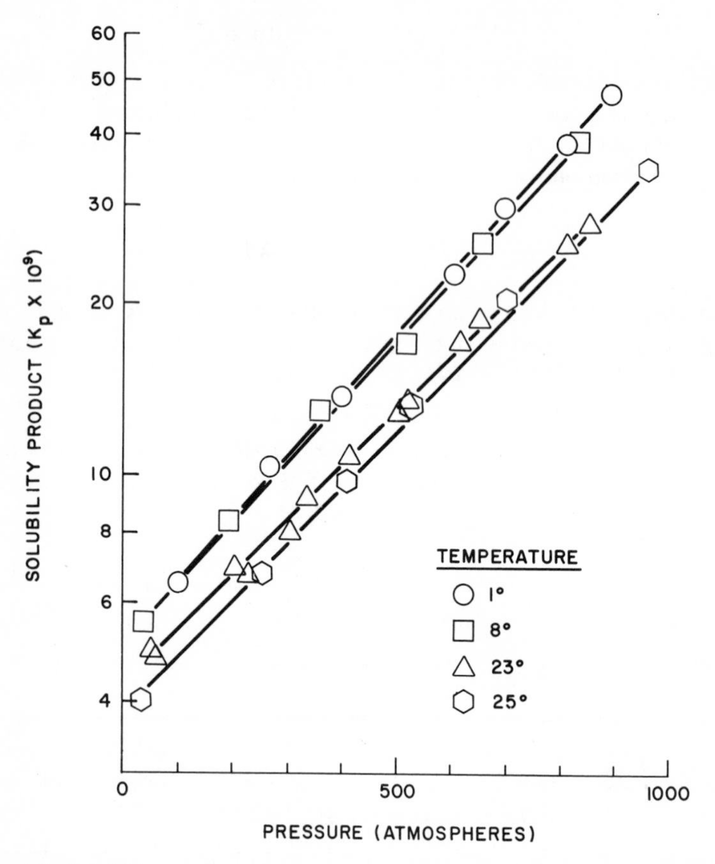

Figure 3.7 Solubility product of calcium carbonate as a function of pressure. Data from MacDonald (1974).

of ΔV have been determined and are compared in Table 3.5 with theoretical values of ΔV.

High Pressures and Temperature

At high pressures and temperatures, the effect of water on the ionization of the solute must be taken into account:

$$MX_{aq} + aH_2O = M^+_{aq} + X^-_{aq} \tag{3.22}$$

for which the equilibrium constant K is

$$K = \frac{K_c}{C^a_{H_2O}} \tag{3.23}$$

Table 3.5 Molal Volume Change for $CaCO_3$ Dissolution

Temperature (°C)	Expermental ΔV (cm^3/g mol)	Theoretical ΔV (cm^3/g mol)
1	-57.1	-63.7
8	-57.5	-62.3
23	-54.9	-59.5
25	-58.0	-59.1

Source. MacDonald, 1974.

Accordingly,

$$\Delta V = -kRT\beta$$

where β is the compressibility of water. Thus from a single experimental value of ΔV, such as in Table 3.5, k can be estimated and ΔV calculated at other temperatures (Marshall, 1972). In applying this technique to solubility phenomena, there could be a problem in knowing the activity or concentration of dissolved molecular salt. However, it has been reported that this concentration may be as high as 20% without affecting the solubility product at 24°C (Doner and Pratt, 1969).

In general, the effect of high pressure and temperature on the solubility product of calcite need not be taken into account since processes for utilization of the brine energy will operate at less than 10 atm and below 300°C. For analysis of reservoir conditions and high pressure processes, however, these effects must be considered.

EFFECT OF CO_2 RELEASE ON CALCITE SOLUBILITY

The release of stripping of CO_2 from the solution as in a flash separator is also important because of its effect, not only on the total CO_2 content, but also because of its effect on pH according to the sequence

$$H^+ + HCO_3^- \rightarrow CO_2(g) + H_2O \quad (3.24)$$

$$HCO_3^- \rightarrow H^+ + CO_3^{2-} \quad (3.7)$$

which shows that carbonate ion will be released. This increase in carbonate concentration will cause calcium carbonate to deposit if the solution is saturated with respect to calcium carbonate. The amount of $CaCO_3$ precipitated, assuming equilibrium conditions, can be calculated from equilibrium and material balance considerations using Equations 3.4 to 3.9. The most convenient form is probably

$$\Delta(\mathrm{Ca}) = \left[\frac{\delta f_2}{\delta(C_T)}\right] \Delta(\mathrm{CO_2}) \tag{3.25}$$

where f_2 is the function relating the equilibrium equations and material balances in the form

$$(\mathrm{Ca}) = f_2(C_T) \tag{3.26}$$

and $\Delta(CO_2)$ is the amount of CO_2 released from the solution. The function f_2 can be obtained from Equations 3.2 and 3.14.

SUMMARY

The quantity of dissolved carbon dioxide present in brine solution can be estimated using Henry's law. The total amount of carbon dioxide species present is the sum of dissolved carbon dioxide, carbonate, bicarbonate, and carbonic acid. The predominate species in the pH range 7 to 8, which is typical of most geothermal brines, is bicarbonate. Flashing of geothermal brine results in the release of CO_2 causing the pH of the solution to increase. This in turn increases the carbonate concentration, which will cause calcium carbonate to deposit if the solution is saturated with respect to calcium carbonate. Data on the equilibrium constant for calcium carbonate over the range of pressure, temperature, and NaCl concentration of geothermal brines can be used to estimate the calcium carbonate deposition that will take place.

NOMENCLATURE

a	arbitrary constant
A	area
C_T	total CO_2 in solution, mol/liter
f	arbitrary function as defined in text

ΔH	enthalpy change, heat of solution for calcite
k_n	reaction rate constant
K_H	Henry's law constant, g-mol/liter-atm
K	equilibrium constant of chemical reaction, see Chapter 2
K_c	solubility product for calcium carbonate
K_{cc}	solubility product for calcium carbonate, corrected for high ionic strengths
K_W	dissociation constant for water
m	molality, mol/kg of solution
MW	molecular weight
R	universal gas constant
(symbol)	concentration of chemical specie whose chemical symbol is given within the parenthesis, mol/liter
W_i	weight percent of component i, g/100 g of solution
z	valence of the ions in solution

Greek Symbols

α	activity
γ	molal activity coefficient
μ	ionic strength

Subscripts

c	calcium
cc	calcium, corrected

REFERENCES

Berner, R. A., "Comparative Dissolution Characteristics of Carbonate Minerals in the Presence and Absence of Aqueous Magnesium Ion," *Amer. J. Sci.*, **265**, 45–70 (1967).

Biscoff, J. L., "Kinetics of Calcite Nucleation: Magnesium Ion Inhibition and Ionic Strength Catalysis," *J. Geophys. Res.*, **73**, 3315–3322 (1968).

Boettcher, A. L., and P. J. Wyllie, "Revision of the Calcite–Argonite Transition, with the Location of a Triple Point between Calcite I, and Calcite II, and Argonite," *Nature*, 792–793 (February 25, 1967).

Doner, H. E., and P. F. Pratt, "Solubility of Calcium Carbonate Precipitated in

Aqueous Solutions of Magnesium and Sulfate Salts," *Soil Sci. Soc. Amer. Proc.*, **33**, 690–693 (1969).

Ellis, A. J., "The Solubility of Calcite in Carbon Dioxide Solutions," *Amer. J. Sci.*, **257**, 354–365 (1959).

Ellis, A. J., and R. M. Golding, "The Solubility of Carbon Dioxide above 100°C in Water and in Sodium Chloride Solutions," *Amer. J. Sci.*, **261**, 47–60 (1963*a*).

Ellis, A. J., "The Solubility of Calcite in Sodium Chloride Solutions at High Temperatures," *Amer. J. Sci.*, **261**, 259–267 (1963*b*).

Girou, M. M., and H. Roques, "Etude des conditions de precipitation at 30°C des differentes varietes allotropiques de $CaCO_3$ en fonction du degre de sursaturation," *C. R. Acad. Sci. Ser. D*, **268**, 1244–1247 (1969).

Helgeson, H. C., and D. H. Kirkham, "Theoretical Prediction of the Thermodynamic Behavior of Aqueous Electrolytes at High Pressures and Temperatures: I Summary of the Thermodynamic/Electrostatic Properties of the Solvent," *Amer. J. Sci.*, **274**, 1089–1198 (1974).

Helgeson, H. C., and D. H. Kirkham, "Theoretical Prediction of the Thermodynamic Behavior of Aqueous Electrolytes at High Pressures and Temperatures: II Debye–Hichel Parameters for Activity Coefficients and Relative Partial Molal Properties," *Amer. J. Sci.*, **274**, 1199–1261 (1974).

Johannes, W., and D. Puhan, "The Calcite–Argonite Transition, Reinvestigated," *Contrib. Mineral. Petrol.*, **31**, 28–38 (1971).

Kirov, G. K., and L. Filizova, "Uber die Maglich–Keiten der Diffusionsverfahren bei der Kristallzuchlung (II)," *Krist. Tech.*, **5**, 387–407 (1970).

Langelier, W. F., et al., "Scale Control In Sea Water Distillation Equipment," *Ind. Eng. Chem.*, **42**, 126–130 (1950).

MacDonald, R. W., and N. A. North, "The Effect of Pressure on the Solubility of $CaCO_3$, CaF_2, and $SrSO_4$ in Water," *Can. J. Chem.*, **52**, 3181–3186 (1974).

Marshall, W. L., "Predictions of the Geochemical Behavior of Aqueous Electrolytes at High Temperatures and Pressures," *Chem. Geol.*, **10**, 59–68 (1972).

Maruscak, A., C. G. J. Baker, and M. A. Bergougnou, "Calcium Carbonate Precipitation in a Continuous Stirred Tank Reactor," *Can. J. Chem. Eng.*, **49**, 819–824 (1971).

Packter, A., "The Precipitation of Calcium Carbonate Powders from Aqueous Solution with Slow Development of Supersaturation," *Krist. Tech.*, **10**, 111–121 (1975).

Weyl, P. K., "The Solution Behavior of Carbonate Materials in Sea Water," *Stud. Trop. Oceanogr.*, **5**, 178–228 (1967).

White, W. B., "The Carbonate Minerals," Chapter 12, in *The Infrared Spectral of Minerals* (V. C. Farmer, Ed.), Mineralogical Society, London, 1974.

CHAPTER 4

Chemistry of Silica as Related to Geothermal Brine Utilization

Silicon dioxide, SiO_2, in the form of silica or silicates constitutes the most abundant chemical specie in the crust of the earth. Consequently, it is found in almost all surface and ground waters. The composition of rocks in the crust varies from 45 to 75wt.% silica, expressed as SiO_2. The lower numbers are generally associated with basalts and peridotites of oceanic and volcanic regions. The higher silica compositions are associated with the rhyolites and granites of the continental regions. However, exceptions to this correlation with oceanic and continental regions are not unusual.

SILICA AND SILICATE STRUCTURES AND OCCURRENCE

The basic structure of both silicas and silicates is a tetrahedral arrangement of four oxygen atoms around one silicon atom as shown schematically in Figure 4.1. The bonding between the silicon and oxygen atoms can be described as being part covalent and part ionic.

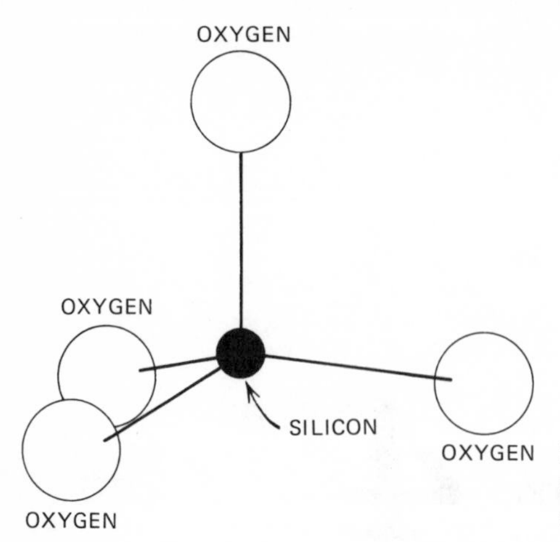

Figure 4.1 The basic tetrahedral arrangement of four oxygen atoms around a central silicon atom.

The silicon atom in the tetrahedral structure is joined to four oxygen atoms by sharing one bond with each so producing an SiO_4^{4-} anion.

Silica

Silica exists in both the crystalline and amorphous states. Examples of amorphous silica are commercially prepared silica gel that contains 20 to 30% water, continuous or gelantinous flocks of gel formed from silica solutions, dissolved or colloidal silica, naturally occurring opal containing less than 12% water, and silica glass prepared by the rapid cooling of silica melts. The various monomeric, polymeric, and colloidal forms of amorphous silica in water solutions are discussed in detail later in this chapter. Silica glass, also called vitreous silica, is an amorphous structure of SiO_2. It consists of an open framework of the Si–O tetrahedron joined corner-to-corner as shown in Figure 4.2. Because many possible ordered arrangements of tetrahedra joined by corners are possible, silica forms various crystal structures, that is polymorphs, as shown in Table 4.1. Because of the slow rate of transition at ordinary temperatures between the various crystalline forms of silica shown in Table 4.1, any of these can exist at room temperature. Stishovite, a high density SiO_6 structure, is a rare exception to the tetrahedral structure of silica and silicates. The solubilities of the different polymorphs of silica are given in Figure 4.3. The solubility of the polymorphs decrease above 300°C. This is not surprising, since at the critical temperature of water and moderate pressure the solubility should approach zero because water is in a

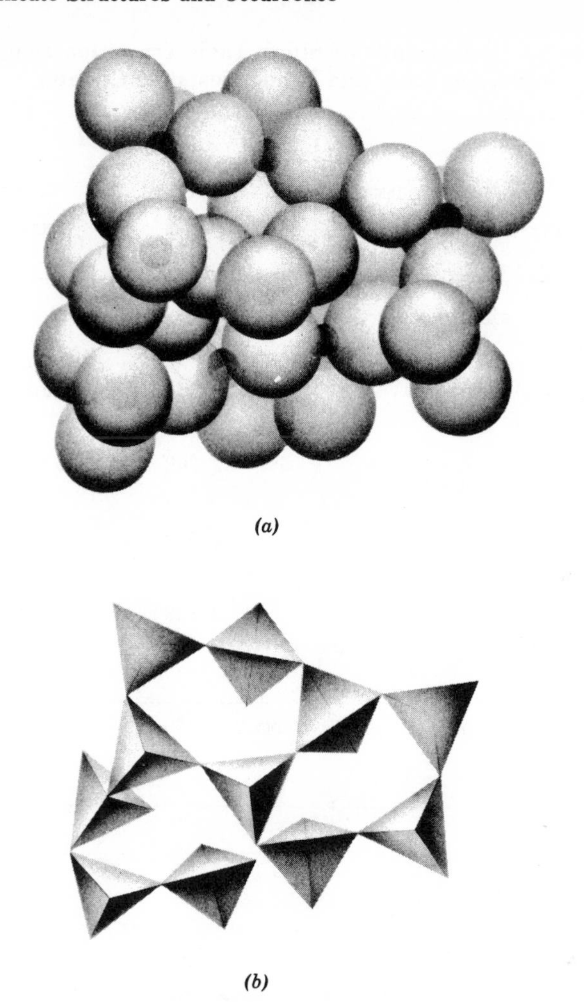

(a)

(b)

Figure 4.2 (*a*) A schematic representation of silica glass in which oxygen atoms are represented by large spheres and silicon atoms by small dark spheres. (*b*) The identical silica glass structure to that in (*a*), but represented in terms of the arrangement of Si–O tetrahedra. The corners of the tetrahedra correspond to the centers of the oxygen atoms in (*a*). From Moffatt et al. (1964).

gaslike state above this temperature. Silica is generally present as quartz, and consequently the concentration of silica found in geothermal brines is determined by the solubility curve of quartz and the temperature of the brine in the underground reservoir.

Equilibrium between quartz and dissolved silica is obtained rapidly at temperatures above 150°C (Morey, 1962). Consequently, the silica

Table 4.1 Polymorphs of Silica, Their Transition Temperatures at 1 atm and Properties at 25°C, 1 atm

Polymorph	Transition Temperature (°C)	Density (g/cm^3)	Crystal Structure
Vitreous silica		---	liquid
	1723		
Cristobalite		2.334	cubic
	1470		
Tridhymite		2.265	hexagonal
	867		
β-quartz		2.648	hexagonal
	473		
α-quartz		2.648	trigonal

Source. Verhoogan, 1970; Othmer, 1969.

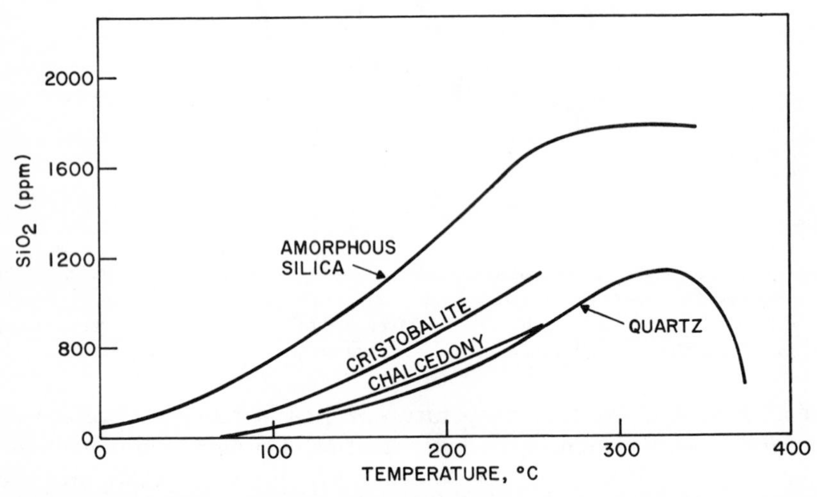

Figure 4.3 Solubility of polymorphs of silica. Quartz is the abundant naturally occuring form and amorphous silica is the form that precipitates from solution at ordinary temperatures and pressures. From Fournier and Rowe (1966).

content of geothermal brine can be estimated from the down-hole temperature using the curve for quartz in Figure 4.3 or

$$\ln K_q^* = 11.2 - \frac{2662}{T} \tag{4.1}$$

Conversely, the silica concentration is often used for estimating the down-hole temperature by assuming that the brine is in equilibrium with quartz at the down-hole temperature. The amount of water lost by evaporation as the brine pressure drops adiabatically to the surface conditions can be determined at either constant entropy or constant enthalpy. Figure 4.4 shows this relationship. The silica method does not always yield estimated temperatures that are in good agreement with measured down-hole temperatures. If cooling of the water by flashing, conduction, or convection causes silica precipitation or if mixing with dilute water occurs, the silica content of the water will be low and the estimated down-hole temperature will be low. Conversely, if the silica content of a hot spring is less than the solubility of amorphous silica at the temperature of the spring, amorphous silica may dissolve from old deposits near the surface and cause an errone-

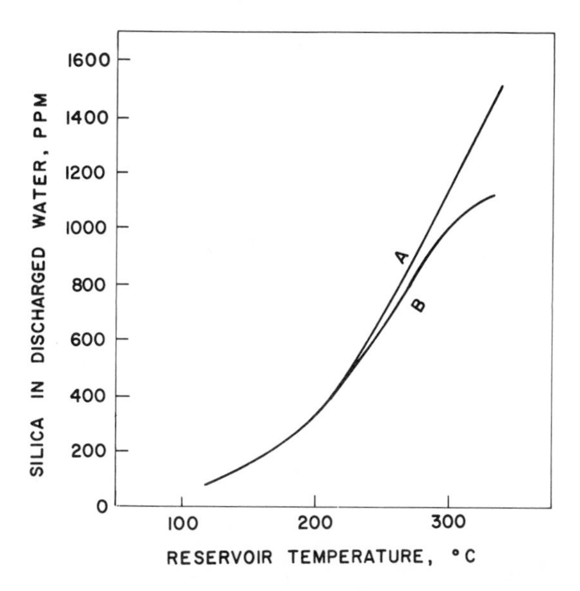

Figure 4.4 Estimation of reservoir temperature from silica content of spring or well brine at the surface assuming equilibrium with quartz at the reservoir temperature. Curve A is for solutions cooled at constant enthalpy, and curve B is for solutions cooled at constant entropy. From Fournier and Rowe (1966).

ously high estimate of the down-hole temperature (Fournier and Rowe, 1966).

Silicates

The corner-to-corner bonds in the framework silica structure can be replaced with cations, thus giving rise to a variety of structures with various oxygen to silicon ratios from 2 to 4 depending on the amount of cations. The single chain structure, Figure 4.5*a*, corresponds to $(SiO_3)^{2-}$, and the double chain, Figure 4.5*b*, to $(Si_4O_{11})^{6-}$. The sheet silicate structure shown in Figure 4.5*c* corresponds to $(SiO_5)^{2-}$.

Silicates occur naturally as different chemical compounds shown in Table 4.2. The simplest are the ortho-silicates in which the silicon atoms are isolated and not linked to one another by the oxygen bridge. Soro-silicates are silicate compounds that contain two silicon atoms joined by one oxygen atom. The smallest ring silicate is three silicon atoms joined by three oxygen atoms. The structure and form of silicates are varied because the oxygen atom can build a bridge between two silicon atoms using any of the four oxygen atoms in the tetrahedral arrangement. In the more complex silicate structures, the 4+ valence silicon atom may be replaced by a 3+ aluminum atom. The proportion is commonly 1 to 4 of aluminum to silica or sometimes 1 to 2. This aluminum substitution results in excess negative charges in

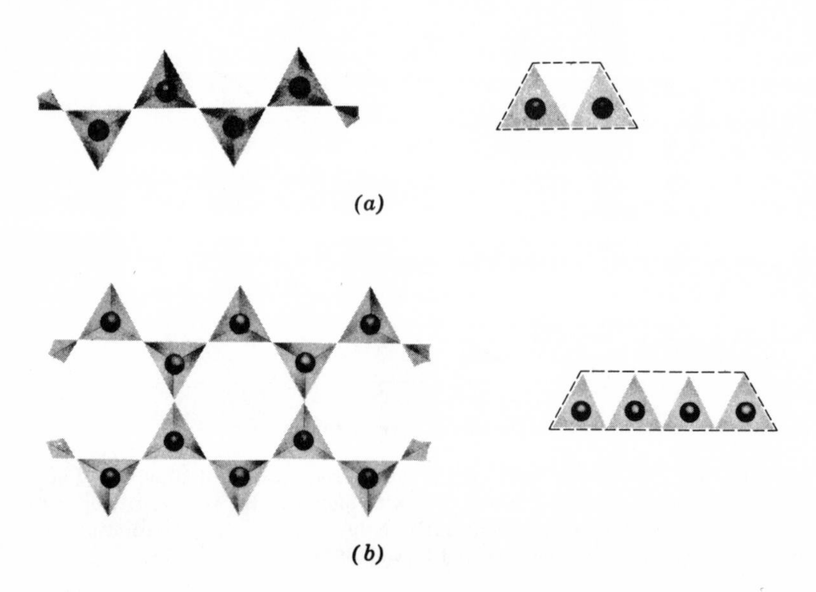

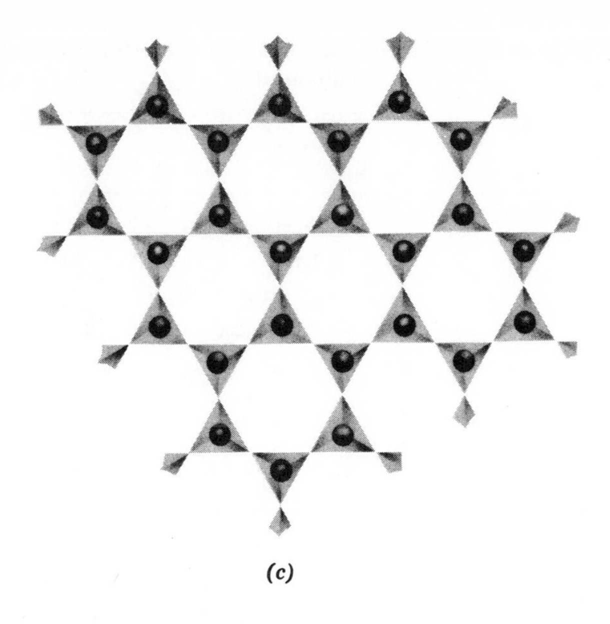

(c)

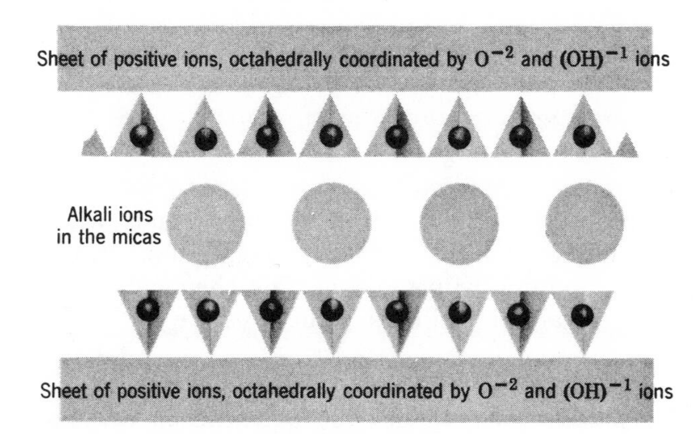

(d)

Figure 4.5 The arrangement of silica tetrahedra in (*a*) one chain of a single-chain silicate in plan view and end view showing the approximately trapezoidal cross section; (*b*) one double chain of a double-chain silicate in plan and end view showing the approximately trapezoidal cross section; (*c*) a sheet silicate; and (*d*) the silicate sheets of a mica crystal viewed on end. The alkali ions actually fit in the large holes shown in the plan view (*c*). In talc and kaolinite, the silicate tetrahedra are arranged similarly in a sheet, but there are no alkali ions holding the sheets together. From Moffatt et al. (1964).

Table 4.2 Silicate Structures and Examples of Naturally Occurring Compounds

Silicate	Formula	Examples[a]
Orthosilicates	SiO_4^{4-}	Zircon: $ZrSiO_4$
		Garnets: $(Ca,Mg,Fe)_3(Al,Fe)_2(SiO_4)_3$
		Olivines: $(Mg,Fe)_2\,SiO_4$
Sorosilicates	$Si_2O_7^{6-}$	$Ca_2Mg(Si_2O_7)$ and $Ca_3Al(Si_2O_7)$
Ring silicates	$(SiO_3)_n^{2n-}$	Beryl: $Be_3Al_2(SiO_3)_6$
Chain silicates	$(SiO_3)_n^{2n-}$	Diopside series: $Ca(Mg,Fe)Si_2O_6$
	$(Si_4O_{11})_n^{2n-}OH$	Actinolite: $Ca_2(Mg,Fe)_5(Si_8O_{22})(OH)_2$
Sheet silicates	$(Si_2O_5)_n^{2n-}OH$	Mica: $KAl_2(AlSi_3O_{10})(OH)_2$
		Kaolinite: $Al_4(Si_4O_{10})(OH)_8$
		Serpentine: $Mg_6(Si_4O_{10})(OH)_8$
Framework silicates	$(SiO_2)_n$	$KAlSi_3O_8$, $NaAlSi_3O_8$, $CaAl_2Si_2O_8$

[a] Cations in parens indicate an and/or combination where the total number of cations within that bracket are given by the subscript. For example, $(Ca,Mg,Fe)_3$ could be Ca_2Fe, Ca_2Mg, Fe_2Mg, etc.

the structure that are neutralized by ionic bonds with other interspersed cations such as calcium, magnesium, iron 2+ or 3+, aluminum 3+, sodium, or potassium. The chemical formula corresponding to these alumino-silicate structures are

$$M(AlSi_3O_8) \qquad \text{or} \qquad M(Al_2Si_2O_8)$$

where M represents one of the above-mentioned cations. No such simple structures exist, however. Rather, the structures consist of the tetrahedral arrangement of oxygen atoms linking the aluminum and silicon atoms.

THE SILICA-SILICATE-WATER SYSTEM

Silica exists in water solutions as mono- or poly-silicic acids. Monomeric silicic acid, H_4SiO_4, is the acid of the ortho-silicate structure. This structure forms higher polymers through oxygen atom linkages. There is no particular upper limit to the molecular weight of these polymers. Colloidal silica will be formed in solution. Silica also forms gelatinous structures. The gelatinous structures can be formed as flocs of silica in a water solution, as a gel structure in a concentrated solution, or as a hard solid of silica gel containing about 20 to 30% water. Because of the relatively slow kinetics of the polymerization and depolymerization reaction at ordinary temperatures, nonequilibrium states of silica will persist for long periods of time. Times as long as several months have been documented.

Equilibria

Dissolved silica is primarily in the form of monomeric silicic acid. This is stoichiometrically the same as two hydrated waters and silica to give ortho-silicic acid, H_4SiO_4. The equilibria involved in the silica–water system are summarized in Table 4.3. As mentioned before, silicic acid polymerizes to form higher polymers of hydrated silica.

The quantity of dissolved silica in solution is determined by the molybdate photometric method. In this method, an acidic ammonium molybdate solution is reacted for a defined time period with the brine water in which the silica concentration is being analyzed. After the reaction is complete, the sample is reduced with sodium sulfite, and the optical density of the resulting blue complex can be determined photometrically. The molybdate reacts within 2 min with the monomeric silica. The higher polymers of silicic acid and colloids in solution do not react in this period of time. Some of the polymeric silica will react if the reaction is allowed to proceed for 10 min. The colloidal silica is essentially unreactive. By this method, the monomeric and the low polymeric silica can be determined in solution.

Solubility. Using the foregoing experimental techniques, the solubility of monomeric silica in equilibrium with amorphous silica as described by Reaction 1 of Table 4.3 has been determined as a function of temperature. The reliable data varies over a 20% range as indicted in Figure 4.6 and follows the normal straight line plot of solubility versus reciprocal temperature. The solid line in Figure 4.6 is an estimate based on the most reliable data. The equilibrium equation corresponding to that line is

$$\ln K_m = -0.86 - \frac{1573}{T} \qquad (4.2a)$$

Table 4.3 Reactions, Equilibria, and Equilibrium Constants for the Silica-Silicate-Water System

Reaction			Equilibrium Constant	
Type	No.	Reaction	Definition	Value at 25°C
Solubility	1	$SiO_2(s) + 2H_2O = H_4SiO_4$	$K_m = H_4SiO_4$	$10^{-2.3}$
Dissociation	2	$H_4SiO_4 = H_3SiO_4^- + H^+$	$K_1 = (H^+)(H_3SiO_4^-)/(H_4SiO_4)$	$10^{-9.7}$
"	3	$H_3SiO_4^- = H_2SiO_4^{2-} + H^+$	$K_2 = (H^+)(H_2SiO_4^{2-})/(H_3SiO_4^-)$	10^{-13}
	4	$H_4SiO_4 + H^+ = H_5SiO_4^+$	$K_3 = (H_5SiO_4^+)/(H^+)(H_4SiO_4)$	---
Polymerization	5	$2H_4SiO_4 = H_6Si_2O_7 + H_2O$	--	
Polymerization	6	$nH_4SiO_4 = H_{2n+2}Si_nO_{2n+1} +$	--	
		$(n-1)H_2O$	--	
Hydrolysis	7	$SiO_4^{2-} + H_2O \rightarrow HSiO_4^- + OH^-$	--	
Hydrolysis	8	$HSiO_4^- + H_2O \rightarrow H_2SiO_4 + OH^-$	--	

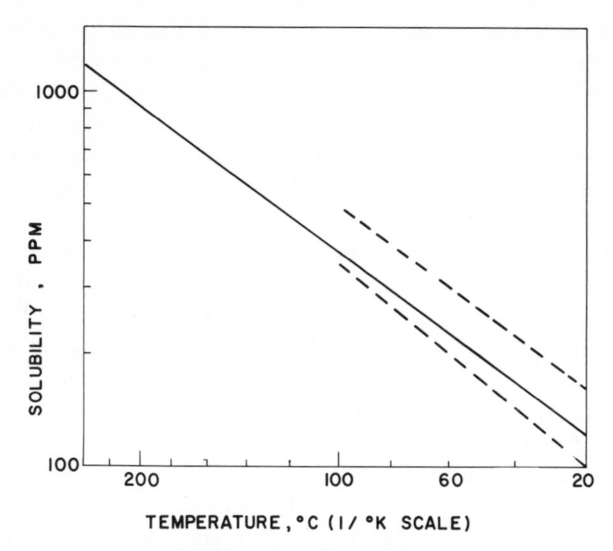

Figure 4.6 The solubility of amorphous monomeric silica. The solid line is based on the compilation of data by Alexander, Heston, and Ihler (1954), Fournier and Rowe (1966), and Krauskopf (1956). The dashed lines indicate the spread of experimental data.

in moles per liter and

$$\ln K_m^* = 10.16 - \frac{1573}{T} \tag{4.2b}$$

in parts per million. Because much data in the literature is given in parts per million, this latter form of the equation for the solubility of silica may be more convenient. In acidic solutions, the ratio of low molecular weight polysilicic acid to monomeric silicic acid is less than 10% for pH values below 11. For very basic solutions above pH of 11, the solubility of silica is greatly increased, the concentration of the polymeric silicic acid decreases, and the concentration of ionized silica increases.

pH and Temperature. The first and second ionization constants of silicic acids are in the range of $10^{-9.5}$ to 10^{-10} and $10^{-11.5}$ to 10^{-15}, respectively. The heat of ionization of silicic acid at infinite dilution has been determined to be 3.3 kcal/g-mol (Greenberg, 1968) over the range 20 to 35°C. Using Equation 2.46, the ionization constant can be calculated at higher temperatures:

$$\ln K_1 = -16.76 - \frac{1661}{T} \tag{4.3}$$

This agrees with the change in K_1 with temperature given by Seward's experimental data (Seward, 1974), although his values at 25°C are lower than indicated by other workers. The heat of solution for amorphous silica is 2.6 kcal/g-mol, which can be obtained from Figure 4.6. Thus the increase in the ionization of silicic acid parallels the solubility of silica. At 240°C the first ionization constant of silicic acid will be about 2×10^{-9}.

Since the first ionization constant is the same order of magnitude as the second ionization constant of carbonic acid, silicic acid ionizes at close to the same pH as bicarbonate ionizes to carbonate. The solubility of amorphous silica in the form of monosilicic acid at 25°C is 0.012wt.% as shown by Figure 4.6 and given by the equilibrium constant K_m in Table 4.3. Since the total solubility of silica w_s in monomeric form is equal to the concentration of molecular silicic acid plus the concentration of silicate ions and, using the equilibrium expressions for Reactions 1 and 2 of Table 4.3, the expression for the relation between pH and solubility expressed in weight percent is given by

$$\log_{10}\left[\frac{(w_s - w_m)}{w_m}\right] = \text{pH} - \log_{10} K_1 \tag{4.4}$$

where w_m is the weight concentration of molecular silicic acid and is equal to 0.012 at 25°C. This equation correctly predicts the solubility of silica in monomeric form as a function of pH as shown by the experimentally determined curve in Figure 4.7. In acidic solution, there is some evidence that the solubility is somewhat increased as denoted by the slight rise in the curve at a pH below 5. An explanation for this is the increase in total monomeric silica due to the specie $H_5SiO_4^+$ according to Reaction 4 in Table 4.3. Note that the accuracy of experimentally determined monomeric silica as indicated by the spread of data in Figure 4.6 covers a range greater than this latter effect. This relationship between pH and solubility of silica has been verified by various laboratory experiments. In addition, a collection of data on silica content of alkaline brines collected from closed basins in Oregon and Kenya shows silica contents varying from 50 up to 2700 ppm for brines in the pH range of 8.6 to 11.0, which generally verifies the pH relationship already discussed (Jones et al., 1967).

Thus the equilibrium relationships for the solubility of amorphous silica to form monomeric silicic acid and for the first ionization constant of silicic acid, Equations 1 and 2 in Table 4.3, enable the total solubility of silica in monomeric form to be calculated as a function of pH. The use of Figure 4.6 or Equation 4.2a to determine w_m, and the

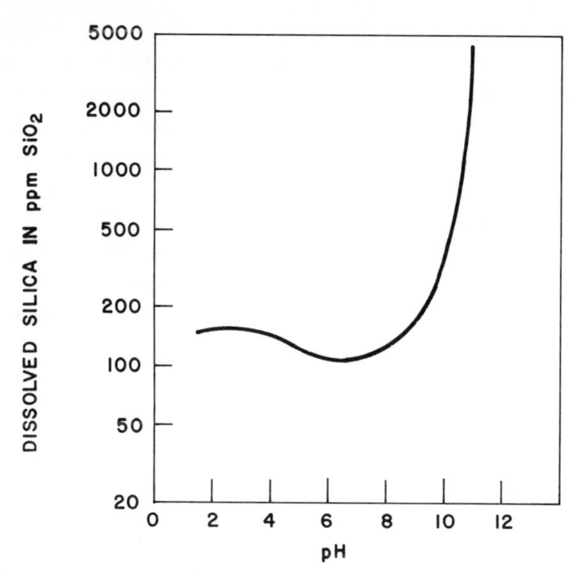

Figure 4.7 The experimentally determined solubility of monomeric silica at 25°C as a function of pH (Alexander, Heston, and Ihler, 1954). Reprinted with permission from *Journal of Physical Chemistry*, Vol. 58, p. 454, Fig. 2, 1954. Copyright by the American Chemical Society.

use of Equation 4.3 for K_1 together with Equation 4.4 determines the solubility of monomeric silica as a function of pH and temperature. Such a prediction has not been compared with experimental data so the accuracy of the method is unknown for higher temperatures.

Polymers. In supersaturated solutions, silicic acid polymerizes with the elimination of water to first form disilicic acid $(HO)_3SiOSi(OH)_3$ and the progressively higher polymers until solutions are formed, as shown by the reactions in Table 4.3. The question then arises, How much silica is present in solution in the form of higher polymers? This question is quite complex, however, because the upper size limit of polymeric silica is an ambiguous quantity since silica forms a continuous molecular weight distribution from monomeric silica on through colloidal silica and including large gel like structures. Methods of predicting the equilibria distribution of polymers are not known.

PRECIPITATION AND/OR POLYMERIZATION

The species that precipitates from solution at ambient pressure is always amorphous silica, whatever the original silica source. The reactions that occur in the silica–water system are demonstrated by the addition of sodium silicate to water. This results in a clear solution and a change in pH towards the basic side, due to the formation of silicic acid by the reaction of water with silicate according to Reactions 7 and 8 of Table 4.3. The amount of sodium silicate that can be added to water at room temperature to form a clear solution is about 500 mg/liter, which is 380 mg/liter greater than the solubility of silica at that temperature. If this clear solution is allowed to stand it will, within a day, form a white precipitate and/or an opaque colloidal solution; if it is heated, it will rapidly form a precipitate at 70°C. This

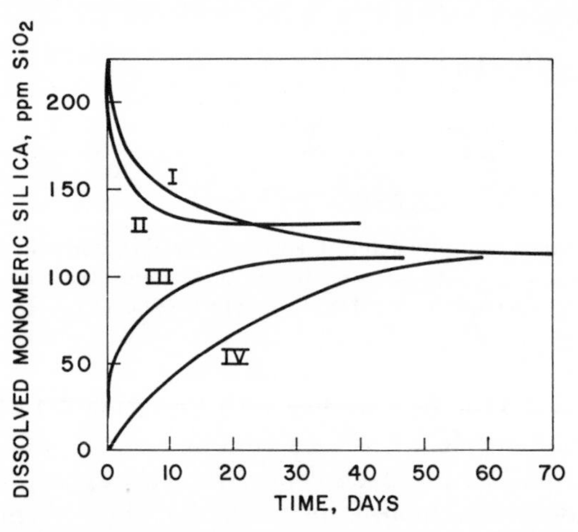

Figure 4.8 Polymerization and dissolution of silica in water at 25°C. In all cases, the equilibrium solubility is approached. Reprinted with permission from K. B. Krauskopf, "Dissolution and Precipitation of Silica at Low Temperatures" in Geochemica et Cosmochimica Acta, © 1956, Pergamon Press Ltd.

Curve	Synthesis	Initial silica (ppm) Total	Initial silica (ppm) Monomeric	pH during run
I	Boiled hot-spring water	320	284	7.7–8.3
II	Na_2SiO_3 neutralized with HCl	975	554	7.3–7.9
III	II, aged and diluted	187	25	8.8–7.4
IV	Silica gel, distilled water	—	—	5.2–5.6

demonstrates that the monomeric silica solution reacts slowly at room temperature to form polymeric and colloidal silica, but reacts rapidly at higher temperatures. Similarly, geothermal brines may flow from the well clear, but after standing for a few hours begin to form a cloudy solution, and later more precipitate forms and settles.

The polymerization of silica can be observed experimentally by measuring the concentration of monomeric silica using the molybdate method in a supersaturated silica solution such as one made with sodium silicate. Likewise, the rate of dissolution can be observed by the same measurements as a function of time starting with unsaturated solutions. Such experiments are shown in Figure 4.8 for the rate of precipitation and rate of solution of silica at 25°C. Figure 4.9 shows some further rate of solution data. As can be seen, several days or weeks may be required to reach equilibrium at this temperature. The greater the degree of initial supersaturation, the faster the concentration of dissolved silica approaches the equilibrium value. This occurs

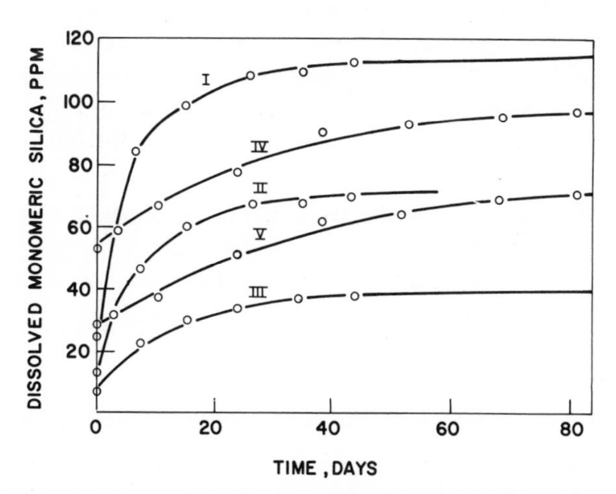

Figure 4.9 Dissolution of silica sols at 25°C. Reprinted with permission from K. B. Krauskopf, "Dissolution and Precipitation of Silica at Low Temperatures" in Geochemica et Cosmochimica Acta, © 1956, Pergamon Press Ltd.

Curve	Total SiO_2 (ppm)	Method of Synthesis
I	183	Aged and diluted Na_2SiO_3 neutralized with HCl
II	92	
III	46	
IV	50	Diluted hot-spring water
V	75	

because of the autocatalytic effect of the already precipitated silica that accelerates further precipitation.

Experimental data indicates the polymerization rate is inversely proportional to the square root of the silicate concentration thus indicating a second-order reaction. However for acidic solutions, the polymerization may be proportional to the fourth power of the concentration of dissolved electrolytes. There is evidence to indicate that the polymer formed in acid solution is different in nature than one formed in basic solution. Furthermore, the precipitation and associated gelation of silicic acid is a function of pH as well as temperature and other ions present.

Nature of the Precipitate and Solution

When sodium silicate is added to water and acidified, it will form a loose, fluffy precipitate, which changes with time. Some of the results of a systematic study of the type of precipitate formed from various brine compositions are shown in Table 4.4. Experiments 1 to 4 show that initially the solution is clear and reaches equilibrium slowly to form a precipitate at silicate concentrations over 220 mg/liter after aging. Heating of the solution, as in Experiments 4 and 7, also causes a precipitate to form as the monomeric silica initially present in

Table 4.4 Silica Precipitation from Brine Solutions Containing 25,000 mg/liter NaCl and 2290 mg/liter KCl

						Initial Precipitate			
Exp. No.	Ca mg/l	Mg mg/l	HCO_3 mg/l	SiO_2 mg/l	pH Initial	@25°C	Heating	@ Boiling	Ppt after Ageing 16 hours at 25°C
1	None	None	None	200	8.0	None	None	None	None
2	None	None	None	300	8.5	None	None	None	Trace
3	None	None	None	400	8.2	None	None	None	Small
4	None	None	None	500	7.2	None	Cloudy @ 190°F	Moderate	--
5	None	None	None	500	8.5	None	None	--	Moderate
6	None	None	500	400	10.3	None	None	None	Trace
7	300	23	None	200	7.2	None	Fluff @ 150°F	Trace	--
8	300	23	None	250	7.2	Trace	Fluff @ 155°F	Small	--
9	300	23	None	400	7.2	Small	Fluff @ 155°F	Moderate	--
10	300	23	None	1500	11.3	Large	--	--	--

solution will form polymeric colloidal silica more rapidly at a higher temperature due to the increased reaction rate. The initial precipitate is a large floc that changes its character after aging. The settling time of the precipitate increases and the volume of the precipitate decreases showing that a finer and more compact precipitate is formed if aged or heated. These latter reactions are very slow and, in general, the form of the precipitate in the ordinary time scale is amorphous silica of a generally light, fluffy nature. The degree of fluffiness of the precipitate appears to be somewhat reversible depending on the pH, temperature, and rate of agitation of the system. Experiments 7 through 10 show that the presence of calcium and magnesium increase the rate and amount of precipitation.

The precipitation of silica from supersaturated solutions is a progressive process in which a colloidal silica is produced that gradually coagulates into a gel or precipitate. The colloidal solution particles will, at sufficiently high concentrations, aggregate to form gels. Common silica gel, widely used as a desiccant, is made by acidifying a solution of sodium silicate and removing the inorganic salts by dialysis. Extensive experimental studies have been made of silica gels. A silica hydrosol formed, for example, by the reaction of sodium silicate solution with hydrochloric acid will, after some time, spontaneously change to the gelatinous, jelly-like mass, silica gel. Gelatination of silicic acid is not accompanied by any noticeable heat evolution. The gelatination takes place essentially by water intrusion into the gel structure combined with a large increase in viscosity. A great part of the water in the initially jelly-like silicic acid may be squeezed out of the framework of the colloid either spontaneously or by external pressure. Dehydration by heating will remove the last traces of water in the gel. Such extensive dissociation is characterized by a series of irreversible processes that alter the properties of the gel.

Effect of pH on Kinetics

The rate of gel formation is strongly influenced by the pH of the solution, as shown in Figure 4.10. The shape of the curve is generated both from experimental data of many workers over a variety of conditions and also from theoretical considerations (An-Pang, 1963). Various workers have found that on the acid side, the pH of maximum stability may range from 2 to 5. "Maximum stability" means the minimum tendency for precipitation from a supersaturated solution of silicic acid. The maximum rate of gelling is within the pH range of 6 to

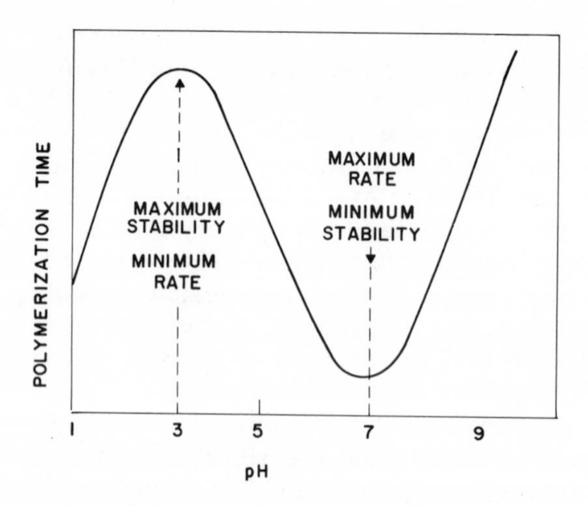

Figure 4.10 Rate of polymerization of silica in aqueous solutions expressed as reaction time as a function of pH based on experimental observations from various sources. The location of the maximum and minimum varies depending on the reaction conditions.

9; the exact point depends on factors such as the concentration of silica and of electrolyte.

In acidic solution, a hydrogen ion will attach itself to the hydroxyl group of silicic acid forming a cationic solution according to equilibrium Reaction 4 of Table 4.3. In acidic solution then, polymerization will take place between silicic acid and the cationic silicic acid to form a dimer or polymer:

$$H_4SiO_4 + H_5SiO_4^+ \rightarrow (H_2SiO_3)_2(H_2O) + H_3O^+ \qquad (4.5)$$

In slightly alkaline or neutral solutions, silicic acid will exist mainly in the form of molecular monomeric silicic acid and a very small concentration of dissociated silicic acid. Polymerization between these two species will occur as follows:

$$H_4SiO_4 + H_3SiO_4^- \rightarrow (H_2SiO_3)_2(H_2O) + OH^- \qquad (4.6)$$

Depending on the equilibrium constant of Reaction 4, which must be considered unknown, concentrations of both cationic and anionic silica may occur in slightly alkaline solutions resulting in a polymerization reaction of the type

$$H_3SiO_4^- + H_5SiO_4^+ \rightarrow (H_2SiO_3)_2(H_2O) + H_2O \qquad (4.7)$$

In highly acidic solution, the concentration of silicic acid will be decreased and the concentration of cationic silica increased thus decreasing the reaction rate. In very acidic solutions of hydrogen chloride, for example, the chloride ion may also attach itself to silicic acid. This will decrease further the concentration of silicic acid that is available for polymerization. In very alkaline solutions, above pH 10, the concentration of anionic silica will significantly increase as exemplified by Figure 4.5 at the expense of silicic acid, thus decreasing the concentration available for reaction. The previously described mechanisms explain the shape of the rate versus pH curve. The maximum rate occurs in the pH range of 6 to 9 where the species for reaction Equations 4.5 to 4.7 are most abundant.

Effect of Temperature on Kinetics

The temperature coefficient of the rate of gelation of silicic acid is dependent on the pH, suggesting that the mechanism of the reaction is different at different pH values. Table 4.5 shows measured activation energies as reported by various workers. As shown by the examples already described, the reaction rate increases significantly so that quite different time scales of polymerization and precipitation are observed between ambient temperatures 25°C and hot brines above 90°C.

The most comprehensive bibliography found in the literature on silicate solubilities and reactions is in OSW R&D Report No. 307 (Collins, 1968). This report includes abstracts from over 200 published sources and covers the following subjects related to silicate reactions:

Table 4.5 Activation Energies for Polymerization of Silicic Acid

pH	Activation Energy (kcal/g mole)	Source
0.7	9.5	Penner, 1946
4.6	16.1	Penner, 1946
5.5	15.5	Brady, 1953
5.6	15.2	Munroe, 1949
8.2	9.4	Munroe, 1949
8.5	9.6	Brady, 1953
10.5	14.6	Brady, 1953

methods of analyzing silicates, buffering in aqueous solutions, and weathering and diagenesis; observations of hydrothermal reactions in nonsaline and saline aqueous media and corrosion prevention. Some data for the solubility of silicate minerals in three-component brine systems are available in OSW R&D Report No. 472 (Collins, 1969).

SILICA-WATER-SALT SYSTEMS

The effect of addition of cations to a silica–water system is in general to increase the rate of precipitation and to either decrease or increase the solubility depending on the cation. Sodium, for example, will increase the solubility of silica in solutions of pH greater than 8. This is due to the formation of the ion pair NaH_3SiO_4 (Seward, 1974) according to the equilibrium reaction

$$Na^+ + H_3SiO_4^- = NaH_3SiO_4 \tag{4.8}$$

This increases the amount of monomeric silica that is in solution because

$$C_s = C_m + C_1 + C_2 \tag{4.9}$$

where specie 1 is $H_3SiO_4^-$, specie 2 is NaH_3SiO_4, and as before, specie m is molecular monomeric silica. The equilibrium constant for Equation 4.8 is 14 at 135°C and 25 at 300°C (Seward, 1974). Combining the equilibrium constant for Reactions 1 and 2 in Table 4.3 and Equation 4.8 with Equation 4.9 gives

$$C_s = K_m \left[1 + \frac{K_1(1 + K_8 C_{\mathrm{Na}})}{C_{\mathrm{H}}} \right] \tag{4.10}$$

where K_8 is the equilibrium constant for Equation 4.8, C_{Na} is the concentration of sodium ion, and C_{H} is the concentration of hydrogen ion. For typical geothermal brines, with pH less than 9 and sodium chloride less than 1 wt.%, the silica solubility is independent of the sodium concentration.

Aluminum and iron are particularly effective in decreasing the solubility of silica. Calcium and magnesium affects solubility as shown in Table 4.4 and discussed previously. Other divalent cations such as strontium and manganese will affect silica precipitation similarly. In general, the insoluble silicates formed from a salt–silicate–water system will be of a complex nature consisting of a distribution of silica hydrates, metal silicates, and hydroxides depending on the pH. Further examples of silica–silicate deposits are given in

Chapter 5. The rate of formation of the insoluble precipitates is influenced by the composition, but the data is limited and generalization is questionable.

Dissolved electrolytes increase the rate of polymerization and coagulation of silica at all values of pH. The electrolytes of seawater, at pH over 7, cause nearly complete precipitation of excess silica in a few hours, regardless of whether the silica is in molecular or colloidal form. Seawater, however, has no effect on the silica in true solution below the equilibrium solubility. Addition of ferric hydroxide causes rapid coagulation of excess silica but does not affect the equilibrium solubility. Since the particles of a silica hydrosol are negatively charged and the hydrosols of iron and aluminum hydroxide are positively charged, a mixture of silica with either of the latter would be expected to coagulate or agglomerate. Aluminum ion precipitates the silica from a solution so that the amount remaining is in the range 2 to 3 ppm for pH between 4 and 5. This is effective in the ratio of 1 part aluminum to 45 parts SiO_2.

SUMMARY

Dissolved silica is in solution as meta- or ortho-silicic acid with a dissociation constant of $10^{-9.7}$. It polymerizes with the elimination of water to form higher polymers and eventually sols. The solubility of amorphous silica increases from about 100 to 1000 pm over the temperature range of interest in geothermal brines.

Silica reacts with divalent cations to form silicates. Iron and aluminum are both known to have reduced solubilities in conjunction with silicate solutions.

The kinetics of silica polymerization are such that silica will precipitate and/or coagulate in a time period from minutes to months depending on the temperature, pressure, pH, and concentration of other cations. The coagulation of silicas is due to a surface effect. The experimental data also indicates that the mechanism of polymerization as well as the type of polymer that is formed depends not only on the pH and temperature but also on the cations and anions that are present. The polymerization rate of mono-silicic acid varies from a gelation time of the order of a few minutes or less to very stable solutions that require days to polymerize. Because of the complexity of the reactions and lack of data and theory and because the time for polymerization spans many orders of magnitudes, the prediction of the reaction rate of silica to form polymers in geothermal brines at various temperatures and salt concentrations is not possible.

NOMENCLATURE

C_i concentration of the *i*th specie, mol/liter

K_i equilibrium constant for the *i*th reaction as defined in Table 4.3, concentration in mol/liter

K_i^* equilibrium constant for the *i*th reaction, ppm

K_m^* solubility product for silica, Reaction 1 of Table 4.3 in ppm and K_m in mol/liter

T absolute temperature, °K

w_i weight concentration of the *i*th specie, parts per hundred = weight percent

Subscripts

i *i*th reaction or specie

m monomeric unionized silicic acid

s total dissolved monomeric silica

q quartz

REFERENCES

Alexander, G. B., "The Polymerization of Monosilicic Acid," *J. Amer. Chem. Soc.*, **76**, 2094–2096 (1954).

Alexander, G. B., W. M. Heston, and R. K. Ihler, "The Solubility of Amorphous Silica in Water," *J. Phys. Chem.*, **58**, 453–455 (1954).

Allen, A. E., and Matijevic, "Stability of Colloidal Silica," *Colloid Interface Sci.*, **31** (3), 287–296 (1969).

Anderson, J. H., and K. A. Wickersheim, "Near Infrared Characterization of Water and Hydroxyl Groups on Silica Surfaces," *Surface Sci.*, **2**, 252–260 (1964).

An-Pang, T., "A Theory of Polymerization of Silicic Acid," *Scientia Sinica*, **XII** (9), 1311–1120 (1963).

Arnosson, S., "Underground Temperatures in Hydrothermal Areas in Iceland as Deduced from the Silica Content of the Thermal Water," *Geothermics*, Special Issue 2, **2** (1), 536–541 (1970).

Barron, L. M., "Thermodynamic Multicomponent Silicate Equilibrium Phase Calculations," *Amer. Miner.*, **57**, 809–823 (1972).

Brady, A. P., G. Brown, and H. Huff, "The Polymerization of Aqueous Potassium Silicate Solutions," *J. Colloid Sci.*, 8, 252–276 (1953).

Brintzinger, W., *Z. Anorg. Allg. Chem.*, **196**, 44–49 (1931).

Collins, A. G., "Silicate Reactions—A Review," OSW R&D Report No. 307, 1968.

Collins, A. G., "Solubilities of Some Silicate Minerals in Saline Waters," OSW R&D Report No. 472, 1969.

Eitel, W., *The Physical Chemistry of the Silicates*, Univ. Chicago Press, Chicago, 1954.

Elmer, T. H., and M. E. Nordbert, "Solubility of Silica in Nitric Acid Solutions," *Amer. Ceram. Soc.*, **41** (12), 517–520 (1958).

Fanning, K. A., and P. O. Pilson, "On the Spectrophotometric Determination of Dissolved Silica in Natural Waters," *Anal. Chem.*, **45**, 136 (1973).

Fournier, R. O., and J. J. Rowe, "Estimation of Underground Temperatures from the Silica Content of Water from Hot Springs and Wet-Steam Wells," *Amer. J. Sci.*, **264**, 685–697 (1966).

Greenberg, S. A., et al., "The Behavior of Polysilicic Acid," *J. Colloid Sci.*, **20**, 20–43 (1965).

Greenberg, S. A., and E. W. Price, "The Solubility of Silica in Solutions of Electrolytes," *J. Phys. Chem.*, **61**, 1539–1541 (1957).

Greenberg, S. A., "The Nature of the Silicate Species in Sodium Silicate Solutions," *J. Amer. Chem. Soc.*, **80**, 6508–6011 (1968).

Iler, R. K., *J. Phys. Chem.*, **56**, 680–683 (1952).

Iler, R. K., *The Colloid Chemistry of Silica and Silicates*, Cornell Univ. Press, New York, 1955.

Jones, B. F., et al., "Silica in Alkaline Brines," *Science*, **158**, 1310–1314 (1967).

Krauskopf, K. B., "Dissolution and Precipitation of Silica at Low Temperatures," *Geochem. Cosmochim. Acta*, **10**, 1–26 (1956).

Mitchell, S. A., "The Surface Properties of Amorphous Silicas," *Chem. Ind.*, 924–933 (June 4, 1966).

Moffatt, W. G., et al., *The Structure and Properties of Materials*, Vol. I, Wiley, New York, 1964.

Morey, G. W., R. O. Fournier, and J. J. Rowe, "The Solubility of Quartz in Water in the Temperature Interval from 25 to 300°C," *Geochim. Cosmochim. Acta*, **26**, 1029–1043 (1962).

Nauman, R. V., and P. Debye, "Light Scattering Investigations of Carefully Filtered Sodium Silicate Solutions," *J. Phys. Chem.*, **55**, 107 (1951).

Penner, S. S., "New Method for Determination of the Activation Energy for the Gelation of Silicic Acid Solutions," *J. Polym. Sci.*, **1**, 441–444 (1946).

Roy, D. M., "Studies in the System $CaO–Al_2O_3–SiO_2–H_2O$; Phase Equilibria in the High-Lime Portion of the System CaO–SiO_2–H_2O," *Amer. Miner.*, 43, 1009–1029 (1958).

Seward, T. M., "Determination of the First Ionization Constant of Silicic Acid from Quartz Solubility in Borate Buffer Solutions to 350°C," *Geochim. Cosmochim. Acta*, **38**, 1651–1664 (1974).

Siever, R., "Silica Solubility, 0–200°C, and the Diagenesis of Siliceous Sediments," *J. Geol.*, **70**, 127–150 (1962).

Strumm, W., et al., "Formation of Polysilicates as Determined by Coagulation Effects," *Environ. Sci. Tech.*, **1**, 221–227 (1961).

CHAPTER 5

Scale Deposition

It is generally recognized that scaling and deposition in process equipment, piping, disposal channels, and reinjection wells is an uncertain problem area for which the solutions are relatively unknown. Consequently, the design of efficient equipment and plants is difficult, and the prediction of economics is unreliable. In present practice, many brines are discarded while still containing useful heat energy, and other brines, with temperatures theoretically more than adequate for commercial utilization, remain untapped because of potential scaling problems. As additional geothermal resources are developed, scaling problems become increasingly important, and an ever greater quantity of potential energy will be wasted until more is known about scale deposition. An understanding of the chemistry of scale deposition from geothermal brines with application of available design data for scale control can effectively minimize costly field trials and improve plant designs.

For design to prevent scaling, the relationship between brine composition, process conditions, and scale depositions is necesary. As described in Chapter 2, composition of brines to be processed by geothermal utilization plants varies from reservoir to reservoir location, the total dissolved solids ranging between 1000 and 400,000 ppm. The lowest temperature geothermal source that can be economically utilized is uncertain at present, whereas the upper temperature limit of geothermal deposits as presently exploited is in the neighborhood of 600°F. Because there are still active volcanoes in the earth's crust, one would expect that ultimately these higher temperature sources will be used.

As hot brine is produced from geothermal wells and flows through the piping system of a geothermal energy utilization plant, there are pressure and temperature changes, boiling phase changes, and turbulent eddies in the liquid stream. In addition, the brine comes in contact with various materials of construction. As a result, the hot brine solution undergoes an intricate chemical and physical time history. In some cases, scale deposition may eventually shut down the system; in other cases, there will be no deposition; and in still others, the deposition may be so great that the system is inoperable. The complexity and consequent difficulty of determining the relationship between brine chemistry, process conditions, and scale deposition is discussed in this chapter.

Calcium carbonate and silica are the dominant constitutents of scales that deposit from geothermal brines. Calcium carbonate is usually found as deposits of calcite, although other material, such as silicates and hydroxides, may be codeposited with it. Cations that form insoluble hydroxides, such as iron or magnesium, may also deposit with the calcium carbonate. Silica may deposit alone as amorphous silica or with other cations in the form of cationic silicates such as iron or aluminum silicates. Beside these, the deposits may include hydroxides of the cations that have coagulated or coprecipitated with amorphous silica. In addition to carbonate, silica, and silicate deposits, hydroxides of cations that form insoluble hydroxides may deposit. A certain amount of water is expected in the scale deposit in the form of chemically bound or adsorbed water plus some occluded water since the deposits are forming in a water environment. Therefore, some soluble salts may be present along with the occluded water in the deposit.

Calcium carbonate solubility increases with decreasing temperature, which is termed retrograde solubility. Silica solubility, however, decreases with decreasing temperature as is usually expected with salts. This is termed prograde solubility. Kinetics of silica deposition are slow, of cationic silicates somewhat faster, while kinetics of calcium carbonate deposition are essentially instantaneous relative to process times. Thus a geothermal brine contains salts that have both prograde and retrograde solubility and both slow and fast kinetics, Table 5.1. Depending on the process conditions, a variety of depositions can be expected.

Table 5.1 Nature of Deposition from Geothermal Brines in Process Equipment

Deposit	Structure	Solubility	Kinetics
$CaCO_3$	crystalline	retrograde	fast
silica	amorphous	prograde	very slow
silicates	amorphous, partially crystalline	prograde	slow

GENERAL DESCRIPTION OF DEPOSITS AND THE PROCESS CONDITIONS THAT FORM THEM

Calcium carbonate is frequently deposited from geothermal brines since calcium and carbonate are often present near their solubility limit. When the pressure is dropped sufficiently, carbon dioxide is evolved from the brine solution increasing the pH and therefore the carbonate concentration, Figure 3.3, thus causing deposition of calcium carbonate. The kinetics of this reaction are very fast so that the deposition will occur at the point, or just immediately after the point, of pressure release that results in gas evolution from the brine. Calcium carbonate deposition can occur in the well casing if the brine flashes in its upward flow from the reservoir. This has occurred for example in a well in the Bolshe-Banny field, U.S.S.R. (Kruykov and Larinov, 1970). The thickness of the calcium carbonate deposit as a function of location in this well together with the temperature profile shown in Figure 5.1 indicates that the deposit is associated with gas release due to pressure drop as the brine flows up the well.

The formation of calcium carbonate deposits is a commonly reported occurrence at wellhead locations where the brine is allowed to flash. In addition, the deposit of calcium carbonate associated with pressure release has been demonstrated in pilot process tests with the East Mesa well 6-1 brines (Wahl, Yen, and Bartel, 1974). The calcium carbonate formation in these tests is a function of the pressure release on the brine. This is described more fully in the next section, which presents a detailed description of these test runs.

Silica will deposit at a rate that depends on the degree of cooling

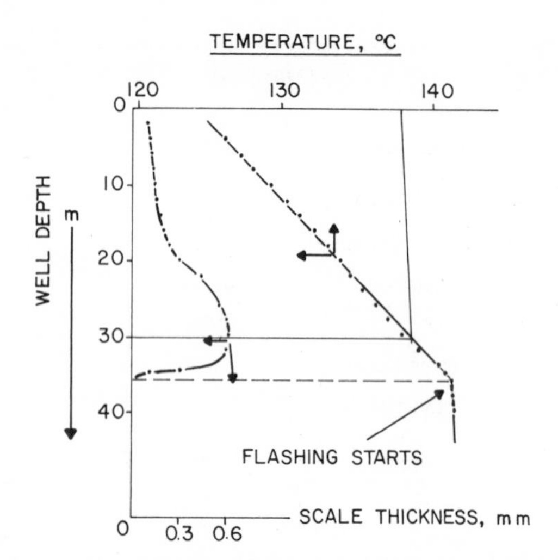

Figure 5.1 Calcium carbonate deposition and temperature versus depth after 6-hr operation. From Kryukov and Larinov (1970). The break in the temperature curve is the location where boiling begins, thus causing a rapid linear temperature drop with decreasing depth and pressure.

below the saturation temperature of the silica. In general, the deposit of silica is relatively slow, taking several hours or more to occur. At Wairakei in New Zealand, silica is deposited mainly in the disposal channels but is also deposited throughout the process equipment (Mahon et al., 1975). The silica deposits are sufficiently voluminous that they must be removed by a power shovel. In Japan (Ozawa and Fujii, 1970), a quiescent period of the brine in hold-up tanks allows the silica to form a colloidal precipitate in the bulk of the solution and thus decreases its tendency to deposit on pipe walls. Deposits of silica from a brine in a heat distribution pipe system in Iceland (Thorhallsson et al., 1975) shows aluminum together with the silica suggesting aluminum silicate. The silica deposits that were formed in the East Mesa tests show a 1:1 ratio of iron to silica, thus suggesting an iron silicate.

A summary of the compositions of scales formed from various geothermal brines under different conditions is shown in Table 5.2. The silica deposits shown are frequently associated with metallic cations. Also note that the calcite may be pure or may deposit simultaneously with silica or silicates. Tests on the deposits of silica and calcium carbonate that codeposited from the East Mesa brines show that the calcium carbonate was formed within a matrix of silica.

Table 5.2 Chemical Composition of Typical Scale Deposits in a Geothermal Process Expressed in Weight Percent

Country	Well	SiO_2	Na	K	Al	Fe	Ti	Ca	Mg
Japan	Otake #1 (Pipe)	78.45	-	-	2.44	2.25	0.20	-	-
"	Otake #2 (Pipe)	87.98	-	-	1.24	0.46	0.04	-	-
"	Otake #3 (Pipe)	93.25	-	-	0.53	0.10	0.06	-	-
"	Matsukawa #1 (Silencer)	51.80	-	-	0.19	9.60	-	0.26	0.23
"	Matsukawa #1 (Separator)	70.80	1.00	0.40	3.77	4.46	-	5.55	1.85
"	Matsukawa #1 (Control Valve)	44.80	8.50	3.50	1.84	2.01	-	0.89	0.33
"	Matsukawa #1 (Exhaust Pipe)	6.00	-	-	Trace	55.85	-	Trace	Trace
"	Matsukawa #1 (Trap)	80.52	2.00	0.80	0.85	6.20	-	0.34	0.11
"	Matsukawa #1 (Wellhead)	17.75	9.25	5.05	0.83	12.20	-	0.21	0.13
"	Matsukawa #2 (Wellhead)	90.45	1.50	0.50	0.84	0.35	-	0.41	0.18
Iceland	Hveragerdi	61.52	0.68	1.05	6.19	2.80	0.25	2.83	0.49
"	Namafjall	73.27	1.06	1.59	5.84	0.18	0.03	2.24	0.23
"	Reykjavik (Steel Pipe)	18.60	-	-	-	47.30	-	0.62	5.78
"	Reykjavik (Galvanized Pipe)	28.00	-	-	-	6.99	-	1.72	0.36
"	Reykjavik (Copper Pipe)	17.10	-	-	-	13.98	-	-	-
"	Reykjavik (Copper Pipe)	30.00	-	-	-	8.39	-	0.71	3.13
U.S.A.	Mesa 6-1 (Control Valve)	0.20	0.12	-	6.40	-	-	34.90	0.10
"	Mesa 6-1 (Separator Entry)	0.94	0.11	0.01	-	2.55	-	33.10	-
"	Mesa 6-1 (Exit Line)	28.20	1.50	-	-	27.20	-	0.87	0.04
"	Salton Sea Sinclair #4 (Exit Line)	36.50	1.40	0.95	-	15.90	-	0.81	0.05
Mexico	Cerro Prieto M-11 (Muffler Waste Trough)	90.02	0.08	0.03	0.07	0.10	-	0.18	0.01
"	Cerro Prieto M-11 (Inside Muffler Area)	85.90	1.60	0.36	0.04	0.08	-	0.31	0.05
"	Cerro Prieto M-6 (Muffler Wall Area)	89.70	0.15	0.08	0.08	0.11	-	0.27	0.02
New Zealand	Wairakei (Pipe Exit)	39.80	0.24	0.01	0.27	5.40	-	0.03	24.20

ell	S	SO_4	CO_3	Zn	Sr	Mn	Pb	Cu	Source
take #1 (Pipe)	-	-	-	-	-	-	-	-	Yanagase, 1970
take #2 (Pipe)	-	-	-	-	-	-	-	-	" "
take #3 (Pipe)	-	-	-	-	-	-	-	-	" "
atsukawa #1 (Silencer)	7.99	1.12	-	-	-	-	-	-	Ozawa, 1970
atsukawa #1 (Separator)	0.25	3.50	-	-	-	-	-	-	" "
atsukawa #1 (Control Valve)	Trace	32.82	-	-	-	-	-	-	" "
atsukawa #1 (Exhaust Pipe)	7.70	2.90	-	-	-	-	-	-	" "
atsukawa #1 (Trap)	Trace	8.80	-	-	-	-	-	-	" "
atsukawa #1 (Wellhead)	3.20	40.84	-	-	-	-	-	-	" "
atsukawa #2 (Wellhead)	Trace	2.25	-	-	-	-	-	-	" "
veragerdi	-	-	-	-	-	-	-	-	Thorhallsson, 1975
amafjall	-	-	-	-	-	-	-	-	" "
eykjavik (Steel Pipe)	-	-	-	-	-	-	-	-	Hermannsson, 1970

Table 5.2 (Continued)

Well	S	SO_4	CO_3	Zn	Sr	Mn	Pb	Cu	Source
Reykjavik (Galvanized Pipe)	-	-	-	41.70	-	-	-	-	" "
Reykjavik (Copper Pipe)	15.10	-	-	-	-	-	-	47.92	" "
Reykjavik (Copper Pipe)	-	-	-	-	-	-	-	39.94	" "
Mesa 6-1 (Control Valve)	-	-	55.74	-	0.60	-	-	-	Wahl, 1974
Mesa 6-1 (Separator Entry)	-	0.21	54.25	-	2.44	0.17	-	-	" "
Mesa 6-1 (Exit Line)	-	-	-	-	0.10	-	-	-	" "
Salton Sea Sinclair #4 (Exit Line)	-	5.59	-	0.19	-	0.73	1.38	6.50	ORC, unpublished data
Cerro Prieto M-11 (Muffler Waste Trough)	-	-	-	-	-	-	-	0.01	"
Cerro Prieto M-11 (Inside Muffler Area)	-	-	-	-	-	-	-	0.01	"
Cerro Prieto M-6 (Muffler Wall Area)	-	-	0.26	-	-	-	-	0.01	"
Wairakei (Pipe Exit)	-	-	0.29	-	-	-	-	0.01	"

This was demonstrated by leaching the calcium carbonate with acid which left a silica matrix.

DETAILED DESCRIPTION OF DEPOSITS FROM A GEOTHERMAL BRINE

An extensive and detailed study of deposition of calcium carbonate, silica, and silicates from East Mesa well 6-1 brines was conducted in connection with the U.S.B.R. project to provide a source of fresh water to supplement the Colorado River. These studies demonstrate the general mechanisms of deposits as well as their chemical and physical characteristics.

Since calcium carbonate deposition is related to pH change caused by carbon dioxide release during flashing, the location of this deposit in the equipment can be related to the location of the pressure drops. Because of calcium carbonate's known retrograde solubility, its deposition is also caused by heated surfaces. Silica and silicate deposition, on the other hand, are related to cooled surfaces and downstream process locations of cooler brine because of its prograde solubility and slow kinetics. These phenomena are demonstrated by the test results that consist of general observations on scaling, chemical analysis of the brine, measurements of scaling in the test rig and associated piping, and detailed analysis of the scale and scaling rates on the test probe and exit line.

capable of simulating a variety of brine processing conditions. The unit was installed at East Mesa well 6-1 as shown in the plot plan, Figure 5.2, and the schematic, Figure 5.3, and was operated from February 16 through April 11, 1974. The retention time of the brine, test section pressure, and temperature varied over wide ranges including vaporization and pressurization above the vaporization point. The flow rate varied from 0.10 to 10 gal/min and so covered a Reynolds number range from laminar to highly turbulent conditions. A temperature loss of less than 3° was incurred through the section of 0.75-in. pipe between the well and the skid. The brine from the skid was discharged into an underground 0.75-in. pipe that connected to an 8-in. dump line above ground. The dump line, at atmospheric pressure, allowed the brine to flow by gravity to the brine pond. Prior to the test operations, the well had been perforated to open a new part of the formation at shallower depth and flowed at a high rate to test its output. After shutdown, the wellhead pressure was zero.

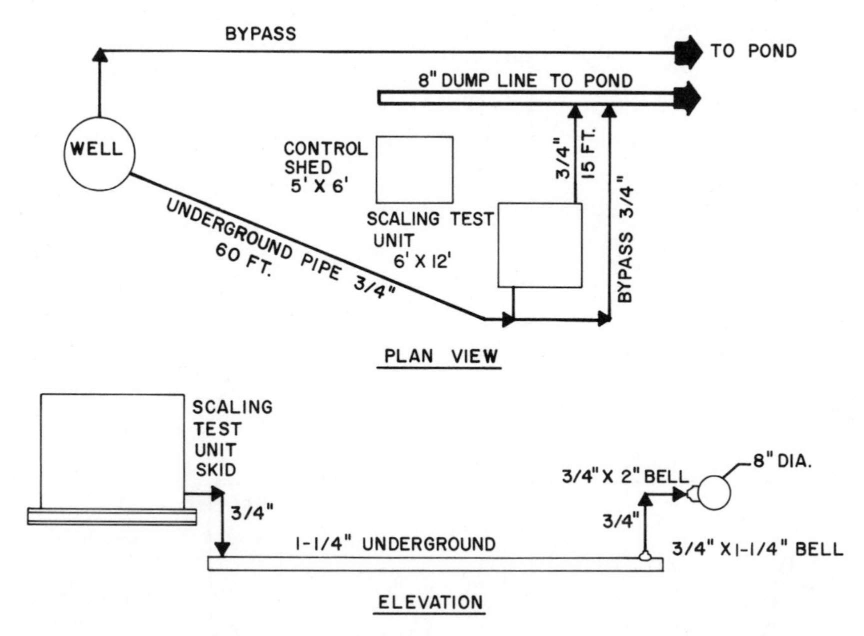

Figure 5.2 Plot plan of a scale deposition test experiment at East Mesa Well 6-1 operated by Occidental Research Corporation under contract to Office of Saline Water, in cooperation with the U.S. Bureau of Reclamation, both under the U.S. Department of the Interior.

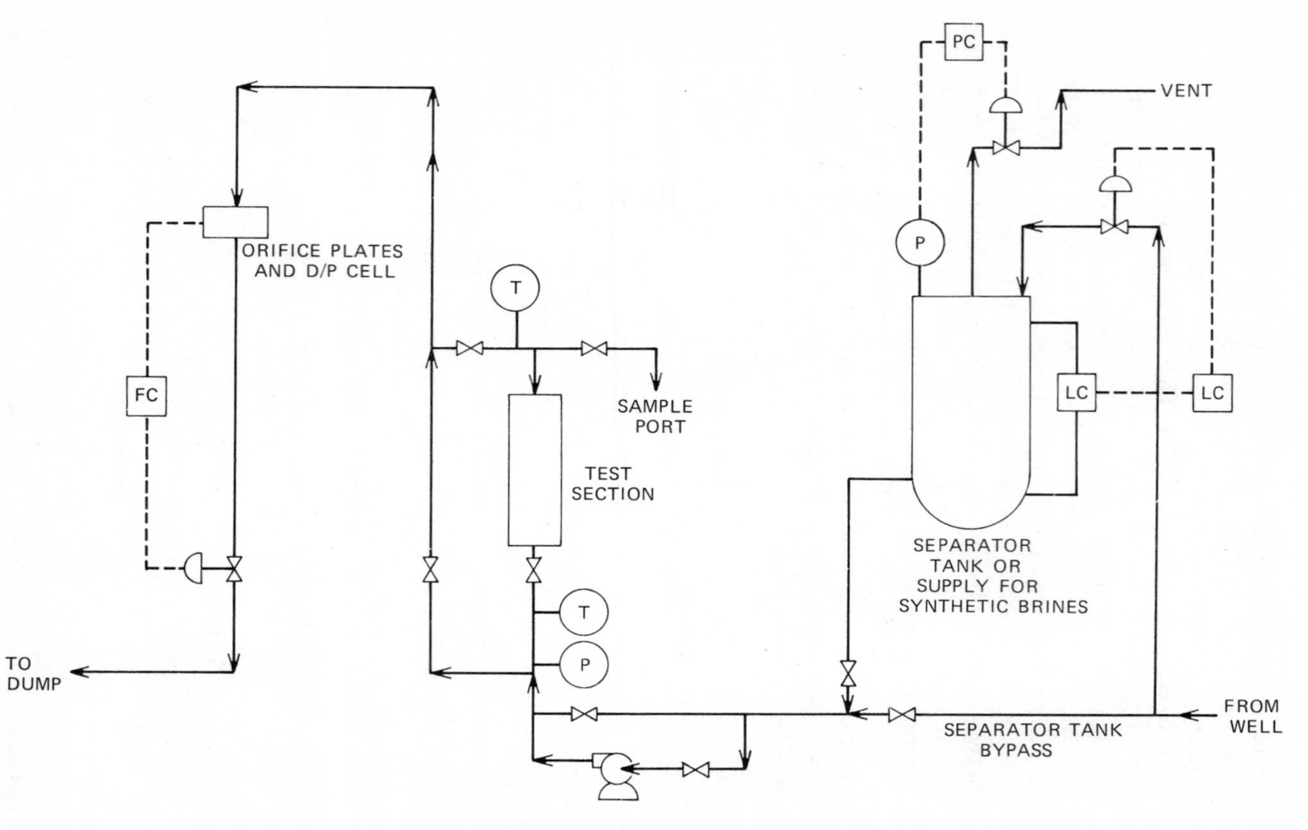

Figure 5.3 Schematic of deposition test unit.

Brine Properties

The well was stimulated by injecting water until the surface liquid pressure was 35 psig. Brine was then produced for 2.5 days prior to testing to flush the injection water from the well. Throughout the test operation the well flowed at 10 to 20 gal/min. During brief shutdowns the well pressure dropped as much as 40%. The well temperature was about 280°F during the runs. Typical composition of the brine during the runs is given in Tables 2.3 and 2.4. The temperature, pressure, and composition fluctuated somewhat but showed a gradual increase during the operating period as shown in Figure 2.2. Ratioing of the brine compositions to a common chloride ion concentration of 14,000 mg/liter, Table 5.3, shows that the ratio of sodium : potassium : silica : calcium : magnesium : chloride : total dissolved solids remains about constant. Also, both the relative and absolute concentration of strontium rose during the first month of operation and appeared to level off during the last month. The concentration of iron varied without any apparent relationship to other variables. There also appears to be a slight tendency for the calcium concentration to rise from a value slightly below one to somewhat above one relative to the chloride ion concentration. No relationship was found between scaling and variations in brine composition other than pH effects that are discussed later. Consequently, the brine description in Tables 2.3 and 2.4 can be considered typical for all the runs. If the pressure fell 20 psi below the well pressure, to about 65 psig, a gas phase began to appear. Since this pressure is above the saturation pressure of water at 280°F, the gas phase occurred because of dissolved gases rather than steam.

Equipment Description

The brine from the well entered the test unit either by way of the separator tank or directly into the test section as shown in the schematic, Figure 5.3. The test section consisted of a 1-in. ID Pyrex tube 18-in. long and a test probe 0.5 in. in diameter and a 24-in. effective length inserted inside the Pyrex pipe as shown in Figure 5.4. Referring to Figure 5.3, it will be seen that the orifice plate used for measuring the flow rate and the control valve for controlling flow rate through the test section are located downstream of the test section, to minimize flashing in the test section. From the control valve, the liquid flows from the test rig through the underground dump line and then to the 8-in. dump line mentioned earlier. A by-pass arrangement

Table 5.3 Composition of Brine from East Mesa Well 6-1 Reduced to a Common Chloride Ion Cconcentration Expressed in mg/liter

Date	Run	#[a]	Day[b]	Time[c]	Well (°F)	Na	K	Cl	SiO_2	Ca	Mg	Fe	Sr	TDS	pH	S[d]	Well psig
2/16	200	1	3	1,390	253	7,499	882	14,000	-	620	17.6	17.6	-	-	-	-	54
2/16	200	2	3	1,390	253	7,509	966	14,000	390	640	17.1	4.3	-	-	-	-	54
2/17	301	1	4	47	275	7,664	891	14,000	-	633	15.8	1.4	-	-	7.90	1	61
2/18	301	2	5	958	278	7,621	875	14,000	-	717	18.6	1.1	-	-	8.30	1	59
2/19	301	3	6	2,430	276	7,315	888	14,000	358	765	18.2	0.3	98	25,735	8.40	1	60
2/23	101	1	9	-	277	5,853	852	14,000	316	813	17.6	0.3	99	24,516	8.20	1	-
2/23	102	1	11	198	282	5,914	879	14,000	313	853	18.5	0.6	104	25,578	7.80	1	66
2/26	103	2	12	2,340	274	-	-	14,000	304	731	14.6	6.7	116	25,480	7.55	1	67
2/27	104	1	13	1,050	262	-	-	14,000	270	746	16.0	6.9	114	25,507	7.90	1	69
2/28	105	1	14	1,400	272	-	-	14,000	261	721	14.3	6.6	117	25,944	8.10	1	66
2/28	302	1	15	74	270	-	-	14,000	282	727	15.5	4.3	115	25,747	8.00	1	69
2/28	302	2	15	450	272	-	-	14,000	292	727	14.4	7.7	117	26,016	6.70	1	70
3/1	302	3	16	1,060	273	-	-	14,000	296	727	14.4	6.6	117	25,602	7.25	1	70
3/1	302	4	16	1,541	270	-	-	14,000	297	739	14.6	4.5	118	26,084	6.70	1	70
3/16	201	1	32	180	269	-	-	14,000	314	744	15.1	0.8	115	23,217	5.75	-	85
3/16	201	2	32	352	271	-	-	14,000	314	677	14.6	1.3	116	23,375	5.75	-	85
3/17	201	3	33	1,167	278	-	-	14,000	242	734	14.8	2.2	124	23,360	5.50	-	85
3/18	201	4	34	2,620	283	-	-	14,000	286	794	15.4	2.0	129	23,955	5.70	-	85
3/19	201	5	35	4,300	283	-	-	14,000	282	750	16.0	3.5	131	22,839	6.00	-	86

3/20	202	1	36	1,413	282	-	-	14,000	274	850	16.5	4.8	142	24,474	5.90	-	86
3/21	202	2	37	2,266	282	-	-	14,000	269	834	17.5	5.0	142	24,397	5.80	-	86
3/22	203	1	38	1,200	282	-	-	14,000	272	854	16.4	8.5	141	24,603	5.80	-	86
3/25	205	1	40	1,270	283	-	-	14,000	-	856	-	4.8	143	24,777	6.30	1	86
3/26	205	2	41	2,605	280	-	-	14,000	-	836	-	3.8	144	24,475	6.30	1	86
3/27	205	3	42	4,215	280	-	-	14,000	275	890	15.3	4.1	126	25,218	6.10	1	87
3/27	205	4	42	4,410	281	-	-	14,000	-	864	-	3.5	134	25,094	6.25	1	86
3/28	206	1	43	968	278	-	-	14,000	256	813	15.5	4.2	137	24,891	6.75	1	87
4/3	207	1	49	120	267	-	-	14,000	291	720	15.7	2.8	125	24,752	6.75	1	89
4/4	207	2	50	1,148	277	-	-	14,000	251	775	16.2	2.7	137	24,924	6.75	1	89
4/5	207	3	51	2,583	278	-	-	14,000	249	793	16.2	2.7	139	24,838	6.70	1	90
4/6	207	4	52	4,120	282	-	-	14,000	250	834	16.1	2.7	136	24,797	7.10	1	91
4/7	303	1	53	1,030	282	-	-	14,000	255	834	16.5	2.5	139	24,789	6.95	1	90
4/9	303	3	55	3,930	282	-	-	14,000	254	812	15.5	2.1	139	25,045	6.60	1	91
AVERAGE	-	-	-	-	275	7,054	890	14,000	286	770	16.0	4.0	126	24,829	6.82	-	78

Source. Wahl, 1974.

[a]Sample Number

[b]Day O is February 13, 1974

[c]From start of run

[d]"1" separator operating, "-" separator bypassed

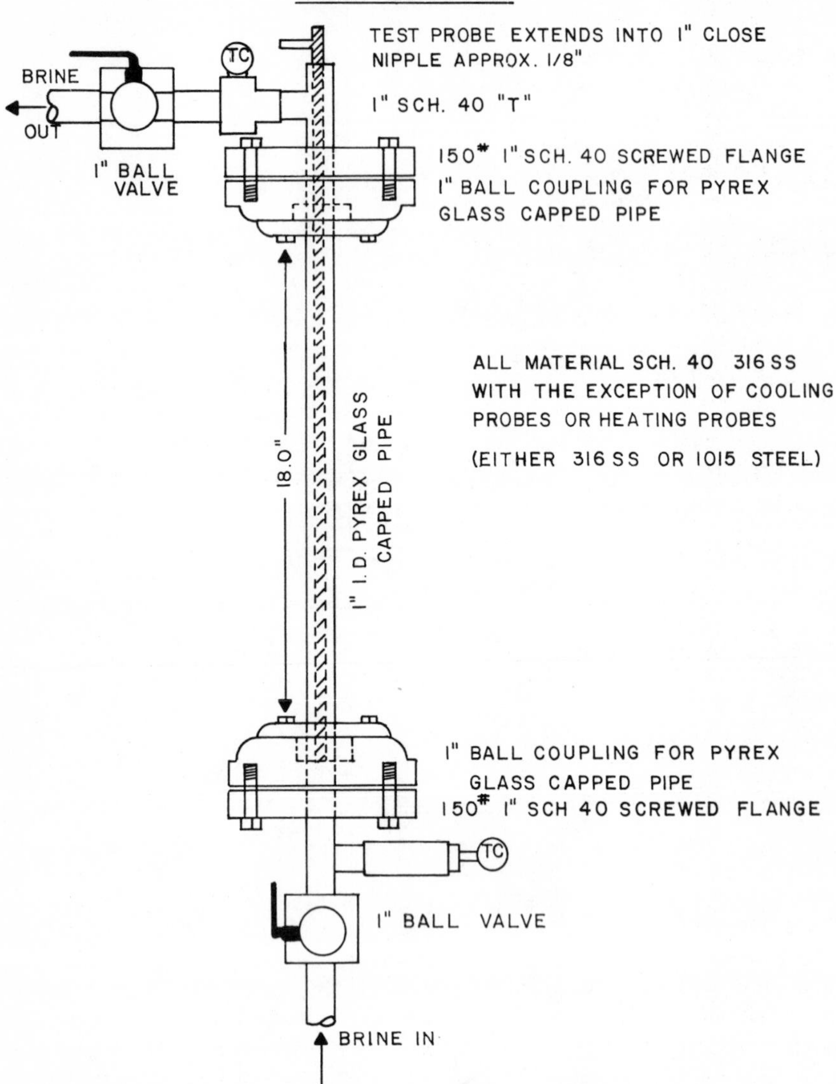

Figure 5.4 Cooled or heated probe test section.

with ball valve shutoffs for the test section was installed so that the system could be operated and brought to its proper operating conditions before the test probe was exposed to the flowing brine. The cooled test probes were cooled with water at temperatures between 100 and 210°F. The electrically heated probes were operated at constant heat flux.

Test Procedure

Throughout the 2-month test period, an effort was made to keep the well flowing continuously. When a short temporary shutdown was required, the well was flowed for an appropriate time prior to restarting the tests to bring the temperature of the brine delivered from the well near its steady-state value. Once the well and separator were operating properly, the test section was opened to the flowing brine and the test run started by opening the inlet and outlet to the test section and closing the test section by-pass valve. At this time, the cooling or heating system was started in operation.

Brine samples were taken daily during the operation, or occasionally more frequently. Samples were taken immediately after the test section so that the brine analysis was representative of the brine composition to which the probe was exposed. Visual observations of the scaling on the probe and at various points in the apparatus were noted. Deposition was observed in the section of pipe following the valve in the exit line.

At the completion of the run, the test section was shut off and the test probe removed. All of the scale that was formed on the probe was removed and the ease of removal noted. The amount of scale was weighed and a visual description of the scale recorded. A summary log, Table 5.4, of all the runs gives the principle operating prameters that were used throughout each run. Scaling of the valve seats and plugs caused the valves to open during the run. Once the valves reached their full open position, the flow rate would begin to fall at which point the test was terminated. The exit line from the test rig where the piping entered the dump line was examined periodically. When the exit line showed significant scaling, the exit line was removed and saved for subsequent examination and replaced with a new section of pipe.

The runs were of three general types: (1) runs numbered 100s, in which the heating probes had no heat flux across the surface, (2) runs numbered 200s made with cooling probes in which the probe surface was cooler than the brine, and (3) runs numbered 300s in which the

Table 5.4 Summary Log of Runs At East Mesa Well 6-1

				Coolant			Test Section			
No.	Day[a]	ΔT Brine to Probe °F	Heater Power Watts	Inlet °F	Outlet °F	Flow gpm	Flow gpm	Inlet °F	Outlet °F	Well Temp. °F
200	3	53	--	188	205	0.9	7.60	250	249	--
300	4	--	1500	--	--	--	2.20	248	250	268
301	4-5	--	600	--	--	--	2.55	249	250	278
				SHUTDOWN						
101	9	--	--	--	--	--	--	240	238	277
102	10-11	--	--	--	--	--	1.20	243	243	278
103	11-13	--	--	--	--	--	1.30	240	240	274
104	14	--	--	--	--	--	1.50	243	243	263
105	15	--	--	--	--	--	1.50	235	235	268
302	15	--	700	--	--	--	2.20	247	247	273
				SHUTDOWN						
201	31-34	68.5	--	190	207	1.30	3.70	269	265	275
202	34-26	104.5	--	154	182	1.40	3.80	276	269	282
203	37	124.0	--	187	109	1.65	1.75	277	265	282
204	38	106.0	--	125	141	1.55	1.65	245	233	267
205	40-42	129.0	--	117	139	1.50	1.75	264	250	283
206	43	117.0	--	109	130	1.45	2.10	243	230	278
207	49-51	132.5	--	116	133	1.45	1.95	243	231	278
303	52-55	--	700-900	--	--	--	1.35	256	258	282

probe surface was hotter than the brine. The heating runs, set up to simulate a vertical tube evaporator, were operated at a flow rate of 1.3 to 2.5 gal/min and a heat flux of 9200 to 20,000 Btu/hr-ft^2. Operation of the separator tank was checked in run 303 by flushing the tank repeatedly during the run to see if there was any effect due to buildup of precipitate in the bottom of the tank. There was no difference caused by such a flushing operation. During Run 200 the pump shown in Figure 5.3 was operated to pressurize the brine flowing through the test unit as well as to recirculate some of the cooled brine. This resulted in less flashing of the brine in the test section and reduced the scaling. The no-heat flux runs used the cooling probes as a test probe and followed the same procedures as the others except no coolant was passed through the test probe.

Table 5.4 (Continued)

	Pressures[b]					Scale		
					Running			
	Well	Separator	Test Sec.	Outlet	time	Total	Rate	Probe
No.	psig	psig	psig	psig	hours	grams	mg/min	Material
200	54	--	68	21	23.2	<0.001	<0.001	316 SS
300	57	24	24	16 to 8	17.8	30.0	30.0	316 SS
301	60	16	17	13 to 5	43.1	32	12.8	316 SS
			S H U T D O W N					
101	--	--	--	--	--	--	--	316 SS
102	66	12	13	9	25.8	1.1	0.7	316 SS
103	67	11	13	8	39.3	<0.02	<0.02	316 SS
104	68	12	13.5	8.4	16.7	0.0	0.0	316 SS
105	66	11	12	5	24.7	<0.01	<0.01	316 SS
302	70	14	15	8 to 10	25.7	21.6	14.0	316 SS
			S H U T D O W N					
201	85	--	60	15 to 36	73.3	0.0	0.0	316 SS
202	85	--	65	26 to 38	37.8	0.0	0.0	316 SS
203	86	--	67	32 to 37	41.3	<0.001	<0.0005	1015
204	84	14	14	2 to 1	14.2	<0.01	<0.01	316 SS
205	87	26	26	12 to 6	75.2	0.04	0.01	1015
206	89	12.5	12.5	0.5 to 1.8	38.6	<0.01	<0.005	316 SS
207	89	12.5	12.2	3.5	69.4	17.8	4.2	1015
303	91	20.5	19.5	6 to 2	65.7[c]	18.0	11.5[c]	316 SS

[a] Day 0 is Feb. 13, 1974.

[b] Prior to day 15, the separator was operated manually rather than on automatic control so there were some fluctuations in early runs.

[c] Scale rate based on last 24 hours of operation because of scale removal during run. See Figure 5.12.

General Observations for a Typical Run

To make the results described later more meaningful, a detailed description of the general observations for run 207 will be given. The exit line was examined prior to startup and showed no scale deposits. The system was started using the by-pass line to isolate the test section for the first 3 hr of operation to allow the temperature of the well to rise and also to allow the separator liquid to reach steady-state composition. Immediately after startup of the test section, the 1013 carbon steel probe showed splotchy black deposits. At 4 hr, it was observed that white flocks deposited on the Pyrex cylinder. At 17 hr, the probe surface was barely visible due to a light-brown scale deposited on the Pyrex tube, and at 23 hr, the probe shape was barely visible. Up to this time, the probe showed no further blackening such as occurred during the first 1-hr operation. At 42 hr, it was observed that the brine was flashing in the test section and the flashing stopped at 45 hr. At 66 hr, the system's temperatures and pressures exhibited a periodic fluctuation with period equal to 5.5 min. The fluctuation was first observed after dark, and the next morning it was observed that heavy flashing was occurring in the test section. The periodic fluctuation may have been due to gas pockets rising through the well. It was observed that the well pressure began to rise at this time. The flow control valve for the brine through the test section was exhibiting indications of scaling because it did not have the range normally associated with its fully open to fully closed operation. Also, the valve would not open completely. This suggests that scaling was occurring. An explanation of this observation also is that the well was exhibiting periodic gas release. Such behavior would make sense in terms of observations noted in previous runs, during which it was frequently observed that after a period of 1 or 2 day's operation the system should exhibit some instabilities and scaling that seemed to occur at a higher rate than during other portions of the runs. The run was shut down at 69 hr as described in the foregoing procedure.

A further example of fluctuation of well performance is demonstrated by run 204. This run was of short duration because fluctuations due to upset of some unknown origin caused the system to behave abnormally. That the well was the source of the variation in behavior is confirmed by observation of the plug in the tee in the entry line to the separator just above the flange at the top of the separator. This plug, Figure 5.5, shows a uniform white deposit with yellow nodules just beginning to form on top of the white deposit. These nodules correspond with fluctuations occurring in the last 1 hr of operation because the plug was new at the beginning of run 204. Another example of variable operation is in run 205. During initial startup of this run, there was some difficulty obtaining a uniform flow. Some blackening of the Pyrex glass and of the probe occurred during the first hour of running time. In the subsequent operation, no further blackening or discoloration of the Pyrex tube occurred at the same rate. Eventually during the latter part of the run, the glass probe became opaque due to a coating on the inside of the Pyrex tube.

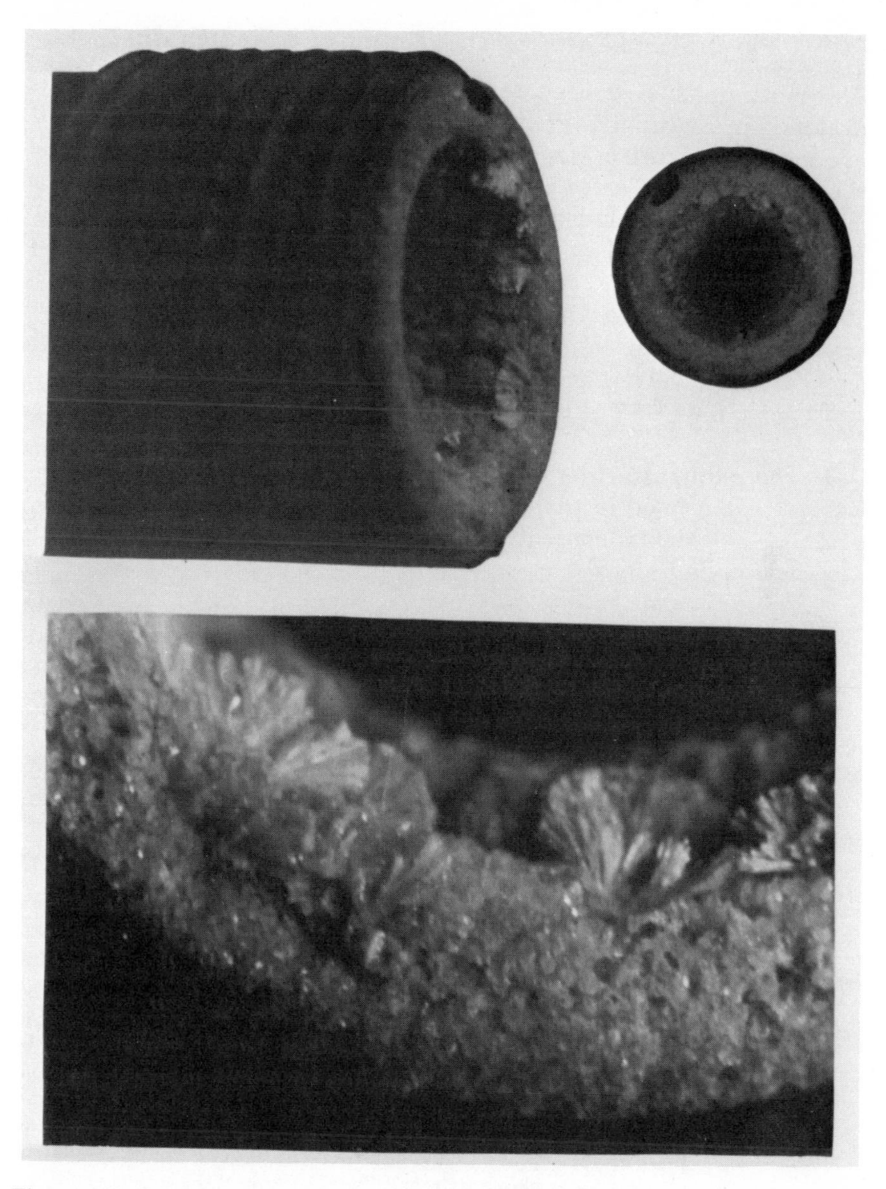

Figure 5.5 Calcium carbonate scale deposits in the pipeline between the well and separator about 1 ft prior to the separator entry and downstream of the flow control valve; see Figure 5.3.

Physical and Chemical Description of Scale Deposits

The most unusual deposit formed was that of run 207, which is sketched in Figure 5.6. Photographs of various sections of the probe are shown in Figure 5.7. A summary of the physical description of the scales is given in Tables 5.5 and the chemical composition in Tables 5.6 and 5.7. Also included in these tables is a description of the scale formed on the control valve during run 205. It was observed that near the end of this run the control valve for the test section would not open its full travel. Subsequent examination of the valves showed that this was due to a thin film of scale less than 0.02-in. thick, which prevented the valve stem from moving into the packing.

Tables 5.6 and 5.7 show the following:

1. The sodium to chloride ratio is approximately 1:1 for all scales except that formed in the exit line for runs 204 and 205.
2. As the calcium content increases, the iron and silica concentration decreases.

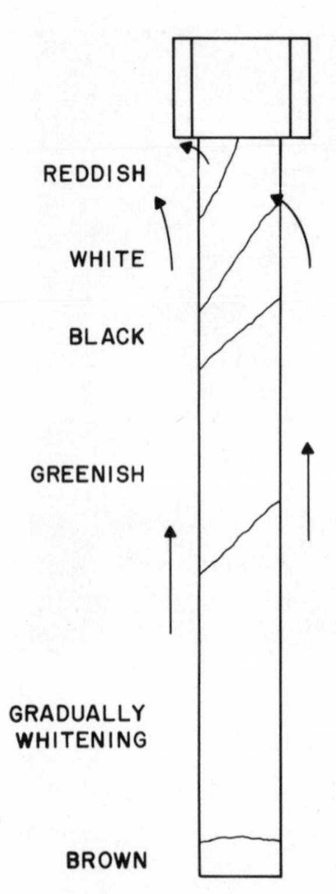

Figure 5.6 Appearance of calcium carbonate scale on a cooled probe from run 207. The colors vary from reddish to white at the top, black on through green in the middle, and black to white to brown at the bottom as shown in this figure. The deposit of white material seems to be predominent on the down stream side of the probe as the fluid went up against the probe and out through the exit line. The scale at the bottom end of the probe is more uniform circumferentially, although there seems to be a slight trend for the color changes to parallel that at the top of the probe. The scale is very hard but the black underlying scale is soft and easily removed. The black underlying scale is very thin. After drying for 1 week, the scale flaked off readily.

Figure 5.7 Scale deposits from run 207. Note the thickness of the scale and the ease with which it flakes off shown in upper two photos. The texture of the scale is shown in the lower left and the different colors from the same scale in the lower right.

3. The iron to silica ratios are approximately constant in all the scales, except for the runs where no silica is present but iron is. Furthermore, the ratio of iron to silica is about 1:1, suggesting that there may be coordination between the two species in the scale buildup.

4. The carbonate is about 20% excess over the calcium. Since iron in solution would form an iron hydroxide rather than an iron carbonate precipitate, and since bicarbonate would not form with the calcium carbonate precipitate, the explanation for the excess carbonate is unknown.

Table 5.5 Physical Description of Scales

Run No.	Sample Location	Appearance	Ease of Removal
101	probe	black	flaked/scraped
102	probe	brown black	scraped
103	probe	brown black	scraped
200	probe	brown powdery scale	scraped off
202	probe	tan film	easily wiped off
203	probe	discoloration like metallic oxidation	easily wiped off
204	probe top	very light black film	easily wiped off
204	probe bottom	greenish brown film	easily wiped off
205	probe	very fine black coating similar to carbon black	easily wiped off
204/5	exit pipe	1/32 in. thick hard glass like black	requires chipping
206	probe top	tan coating, powdery	easily wiped off
206	probe bottom	nodular or crystal build-up plus tan coating	easily wiped off
207	probe	see Figure 5.6	hard & adhering tightly
300	probe	brown black	flakes off
301	probe	brown black	flakes off with light scraping
302	probe	brown black	flakes off with light scraping
303	probe	translucent tan	flakes off readily

5. The excess carbonate is approximately balanced by the iron when there is little or no silica present.

6. Since iron must be balanced by an anion, it is assumed that this is oxygen. The chemical analysis indicated qualitatively that iron was present both in the 2+ and 3+ form.

7. Based on the total weight percentages shown in Table 5.6, there

Table 5.6 Chemical Composition of Scales from Well 6-1 Brine; All Components in Weight Percent

	Run	Na	Cl	CO_3	Ca	Fe	SiO_2	Sr	Mg	Mn	SO_4	K	S	SUM[a]	SUM[b]
Probe	102	0.74	0.81	28.80	14.60	21.00	15.02	0.21	0.31	0.68	-	0.12	0.02	82.31	104.10
Probe	205	5.55	-	-	20.90	6.20	0.86	0.40	0.30	-	-	-	-	-	-
Pyrex pipe	205	0.43	-	56.81	31.50	6.70	0.10	0.40	0.10	-	-	-	-	96.04	100.18
Control valve stem	205	0.12	-	55.74	34.90	6.40	0.20	0.60	0.10	-	-	-	-	96.06	102.08
Probe bottom third	207	0.11	0.26	52.40	29.00	7.23	2.00	0.61	0.11	0.52	0.29	0.03	-	92.56	98.16
Probe top third	207	0.13	0.24	54.18	30.30	6.33	1.20	0.62	0.10	0.52	0.32	0.02	-	93.96	98.53
Separator entry	207	0.11	0.16	54.25	33.10	2.55	0.94	2.44	0.04	0.17	0.21	0.01	-	93.98	96.09
Probe	300	3.34	3.91	47.57	27.80	3.79	6.40	0.39	0.13	-	-	-	0.02	93.35	99.49
Probe, unheated part	301	0.16	0.17	56.33	30.30	8.00	0.24	0.40	0.20	0.55	-	0.01	0.04	96.40	101.41
Probe, heated part	301	0.09	0.11	57.40	30.70	7.62	0.11	0.44	0.20	0.51	-	0.01	0.02	97.21	101.92
Probe	302	0.87	-	11.48	6.11	24.50	29.80	-	0.20	-	-	-	-	72.96	105.73
Probe top	303	0.10	0.01	57.42	33.20	6.40	0.38	0.50	0.20	-	-	-	-	98.21	102.34
Exit line	204/5	1.50	0.02	-	0.87	27.20	28.20	0.10	0.30	-	-	-	-	58.19	91.64
Molecular weight	-	22.99	35.45	60.01	40.08	55.84	60.09	87.62	24.31	54.93	96.06	39.10	32.10	-	-

[a] Of determined components listed in table.

[b] Assuming 2.5 moles OH^- per mole Fe and 2 moles H_2O per mole SiO_2

Table 5.7 Chemical Composition of Scales from Well 6-1 Brine Arranged by Ca Concentration; All Concentration in mol/1000 g

Location	Run	CO_3	Ca	Fe	SiO_2
Exit line	204/5	-	0.22	4.87	4.69
Probe	302	1.91	1.52	4.39	4.96
Probe	102	4.80	3.64	3.76	2.50
Probe	205	-	5.21	1.11	0.14
Probe	300	7.93	6.94	0.68	1.07
Probe, bottom third	207	8.73	7.24	1.29	0.33
Probe, unheated part	301	9.39	7.56	1.43	0.04
Probe, top third	207	9.03	7.56	1.13	0.20
Probe, heated part	301	9.57	7.66	1.36	0.02
Pyrex pipe	205	9.47	7.86	1.20	0.02
Separator entry	207	9.04	8.26	0.46	0.16
Probe top	303	9.57	8.28	1.15	0.06
Control valve stem	205	9.29	8.71	1.15	0.03

is little if any water of hydration present in the scales except perhaps for 2 mol of H_2O, which is necessary to form amorphous silica $Si(OH)_4$.

8. All the other constituents are present in minor quantities with the exception of strontium for the separator entry sample.

Chemical analysis of the scale from the exit line in runs 204 and 205 showed a matrix of silica with a mixture of ferrous and ferric compounds inside. The green and black scale colors observed on the probe are, therefore, probably due to ferro-ferric hydroxides. Chemical analysis of the scales confirms the presence of ferrous and ferric iron. The hydroxide associated with the iron in the scales is somewhere between 2:1 and 3:1, or a value of about 2.5 corresponding to equal amounts of ferrous and ferric ions. Some water may be associated with the silica. Drying the scale from the exit line in runs 204 and 205 at 105°C shows 6.7% water and at 960°C shows 12% water. This later value corresponds to a ratio of 2 mol of water to 1 mol of silica. Assuming that the scale composition consists of a mixture of ferro and ferric hydroxides in the ratio 1:1 and silica with 2 mol of water, the total weight accounted for in each sample is close to 100% as shown in the second sum column of Table 5.6.

In summary, the composition of the scales is a calcium carbonate scale or an iron hydroxide plus silica scale, or some combination of the two. X-ray diffraction analysis shows that the calcium carbonate scales are calcite, and the silica plus iron scales are amorphous.

Pipe Line Scaling

The inlet pressure, or the pressure drop across the exit line for run 201, as shown in Figure 5.8, rises linearly with time. This pressure drop is due to scaling of the exit pipe between the control valve and the 8-in. pipe line. The greatest quantity of scale and the most rapid rate of growth of scale occurred in the section of 0.75-in. pipe just before the 8-in. pipe line. The rate of scaling in the exit line is quantified both in terms of the pressure rise and the amount of scale buildup and is given in Table 5.8. This data shows that the scale buildup is at least ten times as great in the series of runs without the separator in operation as those when the separator was used. Also, the scale formed when the separator was not used was a white calcium carbonate scale, Figure 5.9, whereas the scale formed when the separator was in operation was

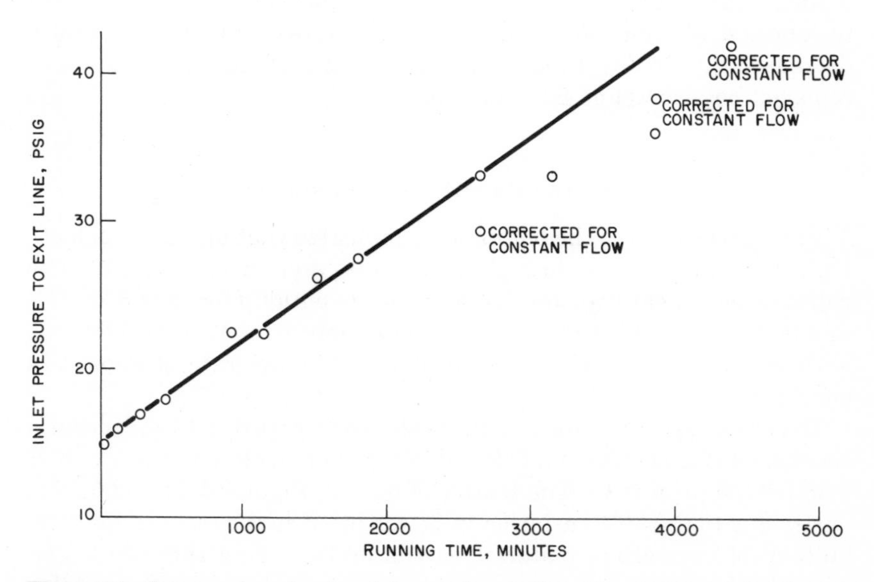

Figure 5.8 Pressure drop due to calcium carbonate deposition in the exit line for run 201.

Table 5.8 Exit Line Scaling

Run	Average Flow Rate (gpm)	Outlet Pressure Initial (psig)	Outlet Pressure Final (psig)	Rate of Resistance Rise $\Delta P/Q^2t$ (10^3psi/gpm^2 min)	Running Time (min)	Scale Thickness (in)	Scale Weight (g/in)
				SEPARATOR TANK BYPASSED			
201	3.65	15	36	.42	4400	0.1	13
202	3.75	26	38	.41	2270	0.2	13
203	3.79	32	37	.19	2485	0.2	13
				SEPARATOR TANK ON LINE			
204	1.7	32	37	∿0	850	0.02	1.5
205	1.7	12	6	descaling	4510	0.02	--
206	2.1	1	1	∿0	2319	0	--
207	2.0	3	3	∿0	1165	0	--
301	2.52	13	2	descaling	2586	0	--
302	1.98	8	9	0	1541	0	--
303	1.33	4	2	0	3940	0.01	1

a black glasslike scale of iron and silica, Figure 5.10. It was observed in runs 204, 205, and 301 that the pressure across the exit line actually decreased due to descaling of the section of exit line which was not replaced between runs 203 and 204.

Heating Runs, Scale Resistance

Visual observation of the boiling rate indicated that uniform nucleate boiling was occurring throughout the test section for run 302. The resistances to heat transfer due to the scale buildup for runs 300, 301, and 302 are shown in Figure 5.11. These approximate a straight line with slope of 0.55°F/kWh, which shows that the rate of deposit is constant.

The scale resistance for run 303 is shown in Figure 5.12 and shows a resistance rise of 0.39 and 0.46°F/kWh, which is close to the value of 0.55 for the prior runs. The scatter of data in Figures 5.11 and 5.12 is partly due to the 10% variation in power input to the test section. The buildup of crystalline material on the probe during this run began with the formation of clear crystalline material at specific nodular sites that eventually formed a 0.03-in. thick coating. The solid

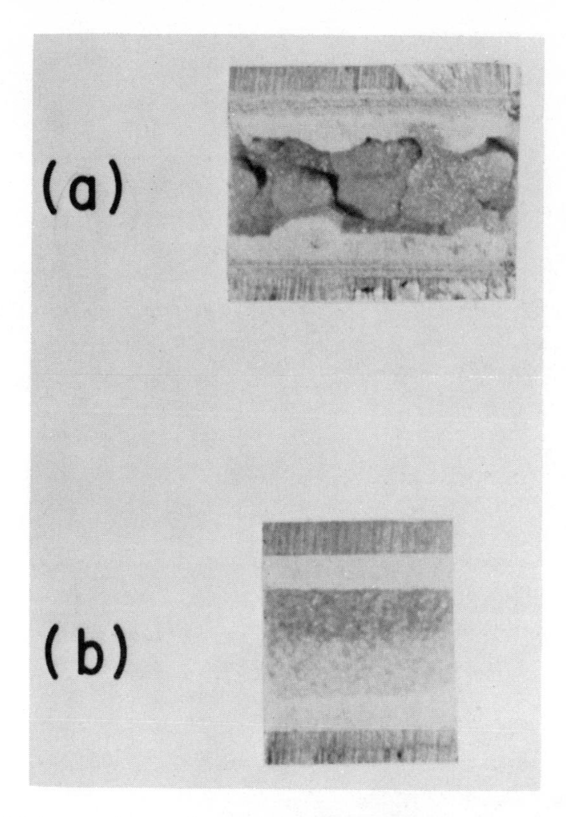

Figure 5.9 Calcium carbonate deposit in a 0.75-in. pipe. Note that the deposit after runs 202 and 203 (*b*) is not uniform in color or thickness, whereas after run 201 (*a*) the deposit was relatively uniform.

material altered the nature of the boiling such that boiling occurred only at a few sites on the probe. During the run, the probe was subjected to a thermal shock by closing the liquid flow to the test section. This caused the temperature of the probe to rise 20°F. During the temperature rise, the boiling of the liquid caused slugs of vapor to move up the test section and the liquid to flow down, which may have increased the thermal shock effect. The result was that a visually estimated 40% of the scale fell free leaving the probe clear and wet. The result on heat transfer resistance is shown in Figure 5.12, which indicates that a larger percentage of the scale was removed than was estimated from a visual observation. However, at the time of the visual observation, the Pyrex pipe was fairly opaque so that it was difficult to estimate the free area.

Figure 5.10 Iron–silica deposit in a 0.75-in. pipe after runs 204 and 205.

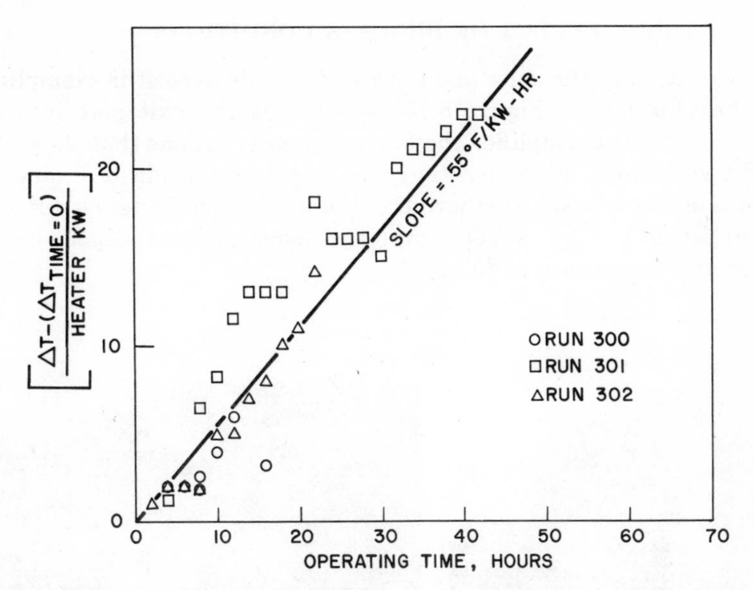

Figure 5.11 Increase in heat transfer resistance at constant heat flux due to calcium carbonate deposition on a heated probe.

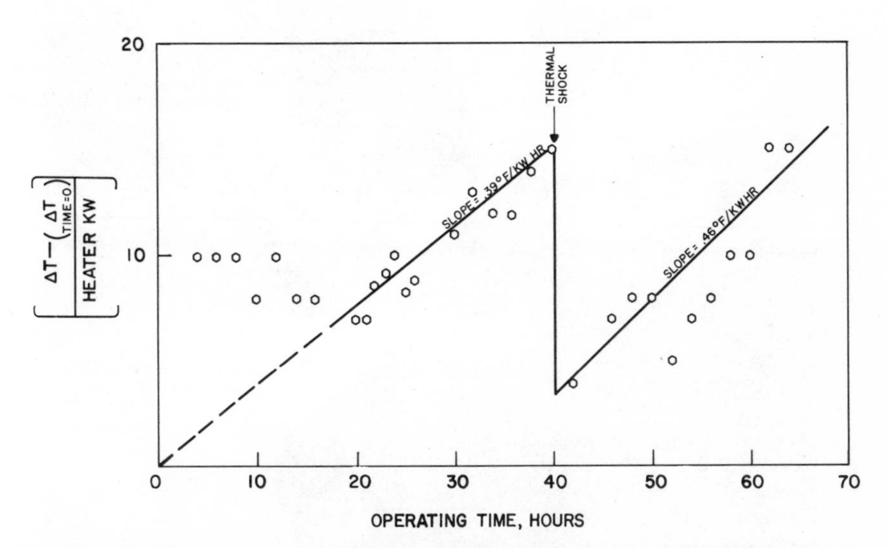

Figure 5.12 Increase in heat transfer resistance at constant heat flux due to calcium carbonate deposition on a heated probe. The thermal shock at 40 hr removed about 75% of the scale resistance.

EFFECT OF PROCESS CONDITIONS

The variation in the color and nature of a scale deposit is exemplified by the pipe elbow, Figure 5.13, which was the exit port into the separator. This exemplifies the foregoing observations that show that slight variations in composition and or process conditions vary the nature of the deposit. Furthermore, the mechanism must be different for apparently slight variations, since sometimes the deposit is uniform and sometimes nodular.

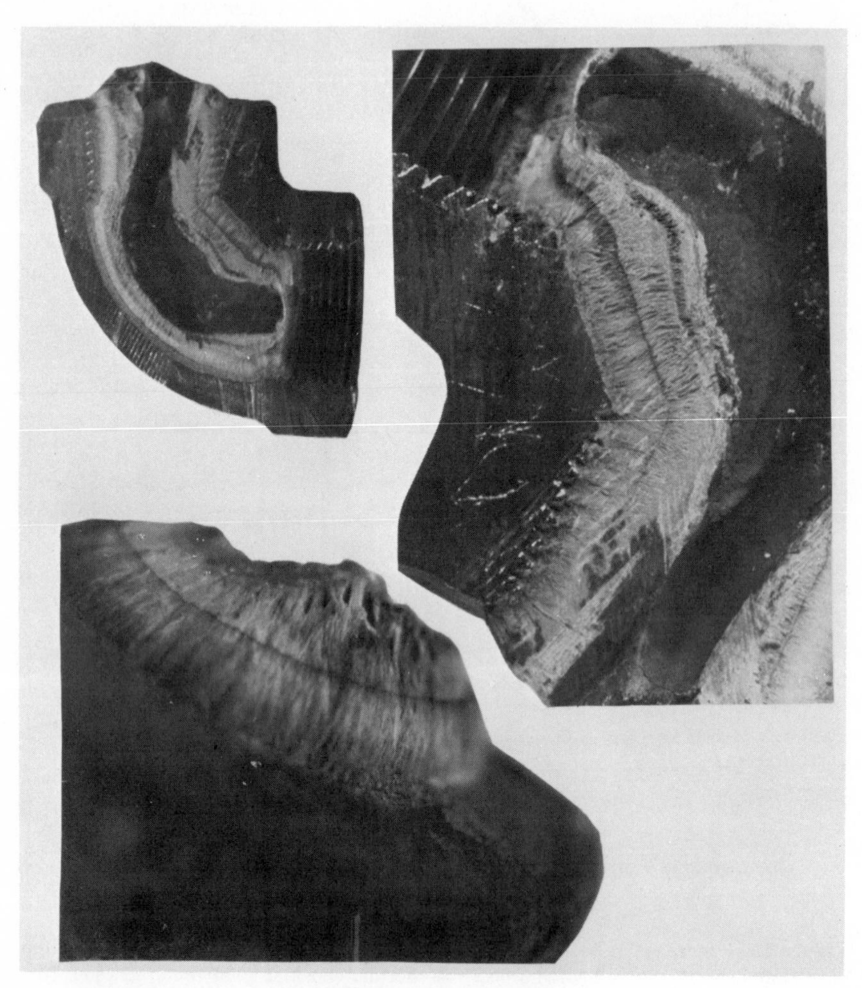

Figure 5.13 Scale in the liquid entry port of the separator after run 105.

Compositions derived from Table 5.6 and the previous discussion are summarized in Table 5.9. Also shown in this table is the scale rate extracted from the log of runs, Table 5.4. The type and location of the scale deposits is related to the flashing of steam and dissolved gases from the brine because of the change in pH resulting from CO_2 release, Equation 3.26. The last line in Table 5.9 shows that for runs 201 through 203 the maintenance of pressure and the prevention of flashing also prevents the deposition of calcium carbonate in the test unit. However at the exit of the scale test unit where the brine flashed into the dump line, calcium carbonate scale formed at a rapid rate. With a slight degree of flashing, runs 205 and 303, the calcite formed on the test probe and an iron–silica scale formed in the exit line at a much lower rate. With a higher degree of flash as shown by the middle lines of Table 5.9, both an iron–silica and a calcite scale formed in the test unit, but the rate of calcium carbonate deposition was somewhat

Table 5.9 Relationship Between Separator Operation and Type, Rate, and Location of Scaling

				Scaling Rate		Type of Scale	
Data Averaged Over Runs	Separator Pressure (psig)	Temp. Drop[a] (°F)	Brine Acidity pH	Probe (mg/m)	Exit Relative[b]	Located On Probe	Located At Exit
206	12.8	31	6.75	0.005[c]	1	$CaCO_3$	--
						SiO_2, Fe[d]	--
103-105	13	26.5	--	<0.01[c]	--	$CaCO_3$	--
						SiO_2, Fe[d]	
302	14.4	24.4	--	14.0[f]	1	$CaCO_3$	--
						SiO_2, Fe[c]	--
303	20.5	23.5	6.78	11.5[f]	1	$CaCO_3$	SiO_2, Fe
205	26	15.5	6.24	0.01	1	$CaCO_3$	SiO_2, Fe
201-203	60[e]	0	5.80	0	20	NONE	$CaCO_3$

[a]Heated Probe.

[b]Based on pressure increase and/or weight of scale deposited.

[c]The low scaling rate in these runs is due to the relatively short duration of the run.

[d]SiO_2, Fe scale could be $(SiO_2.2H_2O)_2Fe(OH)_3$.

[e]Separator was bypassed so that the brine was maintained at full wellhead pressure.

[f]Inlet temperature to separator minus temperature in separator.

less. As the separator operation was increased to give a higher blow down, that is, larger temperature drop, the silica plus iron scale appeared in the test section, and the calcium carbonate scaling rate began to decrease. This shows there is a separator operation or temperature drop that yields a maximum scaling rate at a particular distance or pressure drop downstream of a separator. The high rate of scaling on the heated probe, series 300 runs, compared to the other may be due to the effect of a heated surface on calcium carbonate deposition as described later in the section "Mechanisms and Models for Scale Deposition—Heat Sensitive Deposition."

Comparison of the amount of scale on the probe and the operating conditions for runs 206 and 207 in Table 5.4 confirms the observation mentioned earlier that some undetermined cause related to reservoir or well operation caused a substantial rate of scaling. No significant difference in the analysis of the brine was detected, Table 5.3. Furthermore, examination of Table 5.9 and the composition data in Table 5.7 shows that of the cooled probe runs, run 207 and run 102, both had high scaling rates and were mainly an iron–silicate deposit. This emphasizes the sensitivity of chemical reactions in the brine to small unnoticeable changes in composition that are in an unknown way related to the reservoir system.

As demonstrated by the foregoing example, the prior history of the brine will affect its behavior because it will not be in equilibrium at a given process location. Although the process arrangements that geothermal brine may experience in utilization plants are almost infinite in variety, the number of basic process phenomena are relatively few. These are listed in Table 5.10. Perhaps the most complex process that the brine undergoes is pressure reduction. Pressure reduction can cause flashing of the brine to form a gas phase. This can occur in a process unit in which the gas phase is separated or the two phases may flow together in a pipe system. Another process is that of heat transfer through a surface contacting the brine. This can involve cooling or heating. The resulting temperature change will cause a change in the brine's chemistry and consequently a change in the deposition rates. Associated with any plant will be process equipment with different brine holdup times. Examples are a well-mixed tank, a pipeline in which there is no back mixing, or a settling pond. The effect of pH and chemical additives is a process variation of interest particularly with respect to scale deposition control. The reinjection problem really is a combination of the foregoing processes because it involves flow over surfaces of varying shapes and materials, which may react chemically at varying flow rates, and with temperature change.

Table 5.10 Process Operations That Affect Chemical Properties and Scale Deposition

1. Flashing and pressure reduction with and without separation of the gas phase.
2. Flow conditions, i.e. Reynolds number.
3. Temperature change.
4. Heat flux, cooled, heated or insulated.
5. Holdup time in a process.
6. Holdup time in ponds or tanks.
7. Reinjection.
8. pH.
9. Chemical additives.

MECHANISMS OF AND VARIABLES AFFECTING DEPOSITION

Deposition of solids, either amorphous or crystalline, may occur in the bulk of the solution or on solid surfaces. The former case is referred to as precipitation or crystallization and the later as scale deposition. Deposition occurs first by nucleation followed by growth. Nucleation generally requires a higher degree of supersaturation than growth. If supersaturation is sufficiently high, nucleation will occur in the bulk of a fluid even though no solid particles are present. This is termed homogeneous nucleation. Heterogeneous nucleation occurs on the surface of solid particles such as impurities in solution or may occur spontaneously in the vicinity of already formed crystals. Heterogeneous nucleation will occur on solid surfaces such as pipes, heat exchangers, turbine blades, and so on, due either to the formation of nuclei on a surface infraction or adhesion of nuclei formed in the solution onto the surface. These nuclei then form centers for further growth of scale on the surface. Factors that affect nucleation and growth rate include temperature, concentration of reacting species or degree of supersaturation, and the presence of insoluble impurities such as minute particles.

Depending on the process history, the crystal formation may be colloidal, occur on the surface of the conduit, or form a few larger crystals as compared with colloidal formation. Also, the solution may become supersaturated or unsaturated. One effect of heat flux on scale deposition is the effect of a hotter or colder surface because of the change in the equilibrium constant and of nucleation or crystal growth rate with temperature. Because the change in solubility of calcium carbonate with pH is very large compared to the temperature effect, heat effects will be of secondary importance compared to the pressure changes that cause changes in CO_2 concentration.

The rate of deposition will be the sum of three terms: the formation of nuclei in the bulk of the solution, the formation of nuclei on the surfaces, and the formation of deposits on nuclei already formed either in the solution or on the solid surfaces.

MECHANISMS AND MODELS FOR SCALE DEPOSITION

Deposition of an insoluble precipitate on a surface may occur for a variety of reasons. If the solution is supersaturated, but not sufficiently to spontaneously form new nuclei, the salts will diffuse to a surface where nucleation sites occur and deposit thereon. Deposition may also occur, even if the solution is not supersaturated, if the surface is heated or cooled. Calcium carbonate, for example, will deposit on a hot surface because its solubility decreases with increasing temperature. On the other hand, silica will deposit on a cold surface because its solubility decreases with decreasing temperature. In both of these cases, the deposit of the insoluble salt will be a result of the bulk of the solution containing the salts at a concentration above the solubility of the salt at the surface. In all of the previous cases, the result is that material will be carried from the bulk of the solution to the surface where it deposits. The mechanistic sequence is the transport of material from the bulk of the solution to the surface followed by the chemical reaction of the material to form an insoluble precipitate.

In the case of calcium carbonate deposition, the chemical kinetics are sufficiently fast so that the rate of deposition is limited by the transport of material from the bulk of the solution to the surface. For such a case, the rate of deposition will be proportional to the concentration difference causing the mass transfer. This difference is equal to the concentration in the bulk of the solution less the concentration of the salt at the surface. As is discussed, the effective concentration of the salt at the surface may be the concentration of the salt in equilibrium with the precipitate, or it may be slightly higher if a higher concentration of salt is necessary to cause deposition at a sufficient rate. Mathematically, this is expressed by the relationship

$$R_D = k_D(C_B - C_S^*) \tag{5.1}$$

Silica deposition may be more complex. In the simpler case, silica deposition will be due simply to the transfer of silica from the bulk of the solution to the surface and its precipitation there. If the kinetics of the silica deposition at the surface are sufficiently rapid, then the rate of deposition will be limited by a transfer of the silica from the bulk to the surface according to Equation 5.1. However, if the rate of deposition is limited by the chemical kinetics, that is if the kinetics are slow relative to the transport rate, then the deposition will be proportional to the driving force for the net chemical reaction. The net reaction rate is equal to the forward reaction rate for deposition less the back reaction rate R_S for dissolution:

$$R_D = k_C C_B - R_S \tag{5.2}$$

Since at equilibrium the net reaction rate must be zero,

$$R_D = k_C(C_B - C_S^*) \tag{5.3}$$

Thus the net reaction rate is proportional to the concentration in the bulk of the solution less the concentration in equilibrium with the deposited silica where the constant is the reaction rate constant.

In the more complex case, silica deposition may be due to the formation of polymeric or colloidal silica in solution followed by its transport to the surface and its deposition there. If deposition is limited by the rate of formation of the polymeric or colloidal material, then the mechanism of deposition would be the formation of the deposit on the already-formed silica nuclei at the wall surface rather than in the bulk of the solution. Thus the rate of deposition would be limited by transport to the surface or the reaction at the surface. Since the equilibrium solubility of a solid is zero relative to a fixed surface deposit, the rate of transport or the rate of deposition will be proportional to the concentration of the colloidal material in solution. The concentration of the colloidal suspension will depend on the supersaturation of the bulk solution. Thus the rate of deposition will be proportional to the concentration of the supersaturated solution:

$$R_D = k_D C_B \tag{5.4}$$

The rate of bonding of a colloidal precipitate to a substrate will depend on the forces of attraction, the rate of collision, and the disrupting forces (Wahl and Baker, 1971; Wahl, Pangborn, and McCurdy, 1972). These will depend on the turbulence as determined by the Reynold's number. If the bonding forces are weak relative to the collision forces, then the probability of a successful bond will be low and the rate of deposition will be limited by the bonding rate rather than mass transport so that

$$R_D = k_C C_B \tag{5.5}$$

If both transport of mass and rate of chemical reaction are the same order of magnitude, then the concentration of salt at the surface will be somewhere between the equilibrium solubility and the bulk concentration. At steady state, the two rates will be the same so that

$$k_D(C_B - C_S) = k_C(C_S - C_S^*) \tag{5.6}$$

Eliminating C_S gives

$$R_D = \left[\frac{k_D}{(k_C + k_D)}\right][C_B - C_S^*] \tag{5.7}$$

In all of the foregoing cases, at least in the first approximation according to the aforementioned mechanisms, the rate of deposition is proportional to a concentration difference:

$$R_D = k(C_B - C_S^*) \tag{5.8}$$

For the case of deposition of a colloid at the surface, C_S^* is approximately zero. For very low deposition rates, in which the excess concentration of the species entering into the reaction are low, the reaction rate may decrease substantially due to the importance of nucleation, much as supersaturation is necessary for precipitation in the bulk of the solution. Consequently at lower concentration gradients, a decreasing deposition rate and hence deviation from a straight line relationship would be expected. Also, because of this effect, the value of C_S^* may not be the concentration in equilibrium with solid but may be a higher value.

If deposition is occurring by more than one mechanism, then the deposition rate will be the sum of the individual rates. For example, if silica deposition is occurring due to both the deposit of monomeric silica on the substrate and the deposit of colloidal silica, which is already formed in the bulk of the solution, then

$$R_D = k_1(C_M - C_S^*) + k_2(C_P) \tag{5.9}$$

where k_1 is either a mass transfer or chemical reaction rate constant or a combination thereof as discussed previously for monomeric silica, k_2 is the same factor for polymerized silica, and C_S^* for polymerized silica has been set equal to zero.

The total deposit S of material per unit area at any location in a process plant will be given by

$$S = \int R_D dt \tag{5.10}$$

If the system is operating at steady state, that is process conditions and brine composition delivered from the reservoir are not changing with time, and assuming that the temperature and surface deposit are not changing which should be the case for a no-heat flux surface, then, inserting Equation 5.8 in 5.10 and integrating,

$$S - S_0 = k(C_B - C_S^*)t \tag{5.11}$$

To use the foregoing equations, the rate constant must either be determined experimentally, which is usually the case for chemical reactions, or estimated from known correlations in the case of mass transfer. The mass transfer coefficient is related to the heat transfer coefficient by

$$k_D = \left(\frac{h}{M_D c_P}\right)\left(\frac{c_p \mu}{k_T}\right)^{2/3}\left(\frac{\rho D}{\mu}\right)^{2/3} \tag{5.12}$$

Heat-Sensitive Deposition

Both silica and calcium carbonate deposition will be sensitive to the heat flux. Calcium carbonate deposition will be increased by a heated surface, while silica deposition will be increased by a cooled surface. In either case, the rate of deposition will depend on the concentration of the bulk solution less the concentration of the material in equilibrium with the solid at the temperature of the surface.

In the case of a constant heat flux, the surface temperature of the depositing material will be constant. In this case, the rate of deposition is constant and independent of scale thickness. Consequently, equation 5.11 will again apply. The validity of the equation is shown by the linear increase in heat transfer resistance at constant heat flux for the deposition of calcium carbonate as shown in Figures 5.11 and 5.12. This linear relationship for calcium carbonate deposition is also demonstrated by the data of other workers as shown in Figure 5.14. According to this figure, the concentration difference causing mass transfer and deposition must be different than the concentration between the bulk phase and that in equilibrium at the temperature of the surface. Apparently some different temperature or concentration represents the value of the deposition at low deposit rates, which as

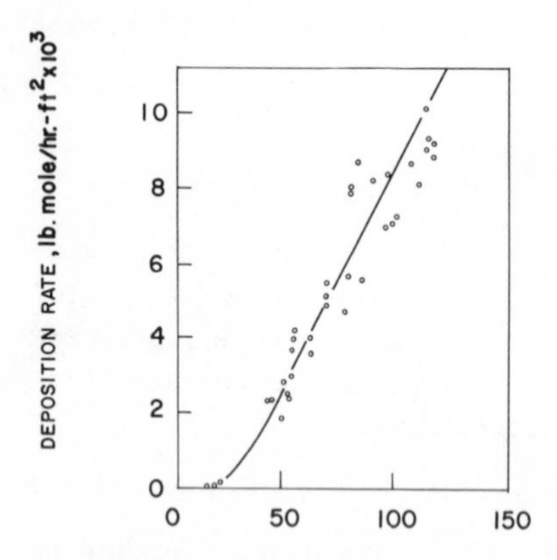

Figure 5.14 Deposition rate of calcium carbonate on a heated surface as a function of the concentration difference between the bulk solution and the concentration in equilibrium with solid calcium carbonate at the temperature of the interface. Concentration is expressed as parts of calcium per million parts solution. The data is from various workers as compiled by Hasson et al. (1968) and reduced to a common basis of 25.5 cm/s. The line through the data differs from that drawn by Hasson. Reprinted with permission from *I&EC Fundamentals*, Vol. 7, p. 65, Fig. 11, 1968. Copyright by the American Chemical Society.

mentioned earlier is to be expected for a crystallization phenomena at a surface. The equation of the straight line in Figure 5.14 is

$$R_D = 1.1(w_B - w_S^*) \tag{5.13}$$

The constant of proportionality for a mass transfer limited process should be given by Equation 5.11. The analogy between heat and mass transfer, Equation 5.12, together with experimental data on heat transfer (Hasson et al., 1968) showed that the constant should be proportional to the 0.72 power of the Reynolds number. This compares with the experimentally determined value of 0.68, showing that agreement between the foregoing mechanisms, mathematical equations, and experiment are satisfactory.

A constant flux heat exchanger is not as common in practice as a heat exchanger operating with a constant temperature cooling or heating fluid. In such an exchanger, the buildup of scale causes an increase in heat transfer resistance and decrease in heat flux for the approximately constant temperature difference between heated and cooled surfaces or fluids.

In the case of constant temperature fluids flowing past the cooling or heating surface, the rate of deposition of material will vary with time because the scale thickness will cause a decrease in heat flux and a corresponding change in the temperature of the surface. This is because the rate of deposition depends on the concentration in equilibrium with the solid, and this is a function of temperature. Over a limited range, the equilibrium concentration is in direct proportion to the temperature of the surface:

$$C_S^* = a_1 T_S \tag{5.14}$$

This can then be used in any of Equations 5.1 through 5.9 and inserted in Equation 5.10 to determine scale thickness as a function of time. As an example, the application of this procedure to the interpretation of experimental results on silica deposition is described later.

Silica scale deposition on a cooled surface was studied using a modification of the apparatus shown in Figures 5.3 and 5.4. The synthetic brine used, Table 5.11, is similar to the brine from East Mesa well 6-1. The silica content was sufficiently high so that polymerized and colloidal silica was present. Under these conditions, the amount of monomeric silica is constant at the saturation value and Equations 5.11 and 5.12 apply so that

$$R_D = k_1(C_M - a_1 T_S) + k_2(C_p) \tag{5.15}$$

Table 5.11 Description of Synthetic Geothermal Brine

Component		Compound Used for Synthesis	
Name	(mg/l)	Name	(g/l)
Na	10,000	NaCl	25.43
K	1,200	KCl	2.29
Ca	300	$CaCl_2 \cdot 2H_2O$	1.10
Mg	23	$MgCl_2 \cdot 6H_2O$	0.193
Li	15	LiCl	0.091
Br	10	$NaBrO_3$	0.019
SO_4	10	Na_2SO_4	0.015
B	20	H_3BO_3	0.115
HCO_3	500	$NaHCO_3$	0.687
SiO_2	500	$Na_2SiO_3 \cdot 9H_2O$	2.360
Cl	17,150	Total Chloride From Above	

For a given run in which flow rates and temperatures are constant,

$$R_D = a_2 - a_3 T_S \tag{5.16}$$

where a_2 and a_3 are constants as defined by the foregoing equations because both the monomeric and polymeric silica concentrations are constant. If k_2 varies with temperature, then that effect would be included in the a_3T_3 term. A schematic representation of the heat flow and heat transfer resistances is shown in Figure 5.15. Since the heat flux is equal through all of the resistances at steady state, the surface temperature of the scale can be related to the brine temperature and coolant temperature by

$$q = \frac{(T_B - T_C)}{(H_1 + H_2 + H_S)} = \frac{(T_B - T_S)}{H_2} \tag{5.17}$$

where the terms are defined in Figure 5.15.

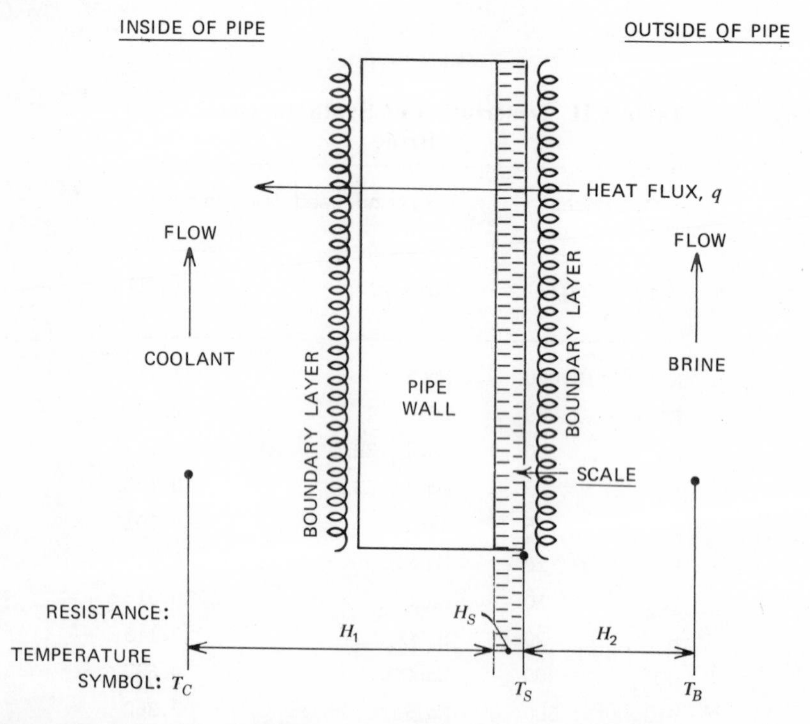

Figure 5.15 Schematic model for scale deposition on a cooled probe. The heat transfer resistances and temperatures used to derive Equation 5.24 are shown on the lower portion of the schematic.

From Equation 5.17

$$T_S = T_B - \frac{H_2(T_B - T_C)}{(H_1 + H_2 + H_S)} \tag{5.18}$$

Since the scale thickness is proportional to the heat transfer resistance and using the differential form of Equation 5.10,

$$R_D = \frac{dS}{dt} \propto \frac{dH_S}{dt} \tag{5.19}$$

Substituting Equation 5.19 in Equation 5.16 gives

$$\frac{dH_S}{dt} = a_2 + a_3 T_S \tag{5.20}$$

Combining this with Equation 5.18 and grouping all constants together

$$\frac{dH_S}{dt} - H_S = \frac{K}{\theta} \tag{5.21}$$

where $\theta = (H_1 + H_2 + H_S)/(a_2 + a_3 + T_S)$ and K is a constant. During initial scale formation, H_S is negligible compared to H_1 and H_2, so that θ can be assumed constant. With this assumption, the solution to Equation 5.21 is

$$H_S = K(1 - e^{-t/\theta}) \tag{5.22}$$

In this equation, the constant K will be the value of the scale resistance that is built up after a long period of time. The time constant θ for the equation is the time it takes the scale resistance to reach $1/e$ of its final value. The ratio of these two parameters is the initial rate of scale formation as expressed by

$$\frac{dH_S}{dt}(\text{at } t = 0) = \frac{K}{\theta} \tag{5.23}$$

Since three characteristics of the rate of scale formation are represented, the ratio K/θ is a representative number useful for classifying the tendency of scaling for different process conditions. Equation 5.22 can be put in dimensionless form so that data from different runs will fall on one curve:

$$\left(\frac{H_s}{K}\right) = 1 - e^{-t_r} \tag{5.24}$$

where $t_r = t/\theta$. Experimental runs at brine concentrations varying

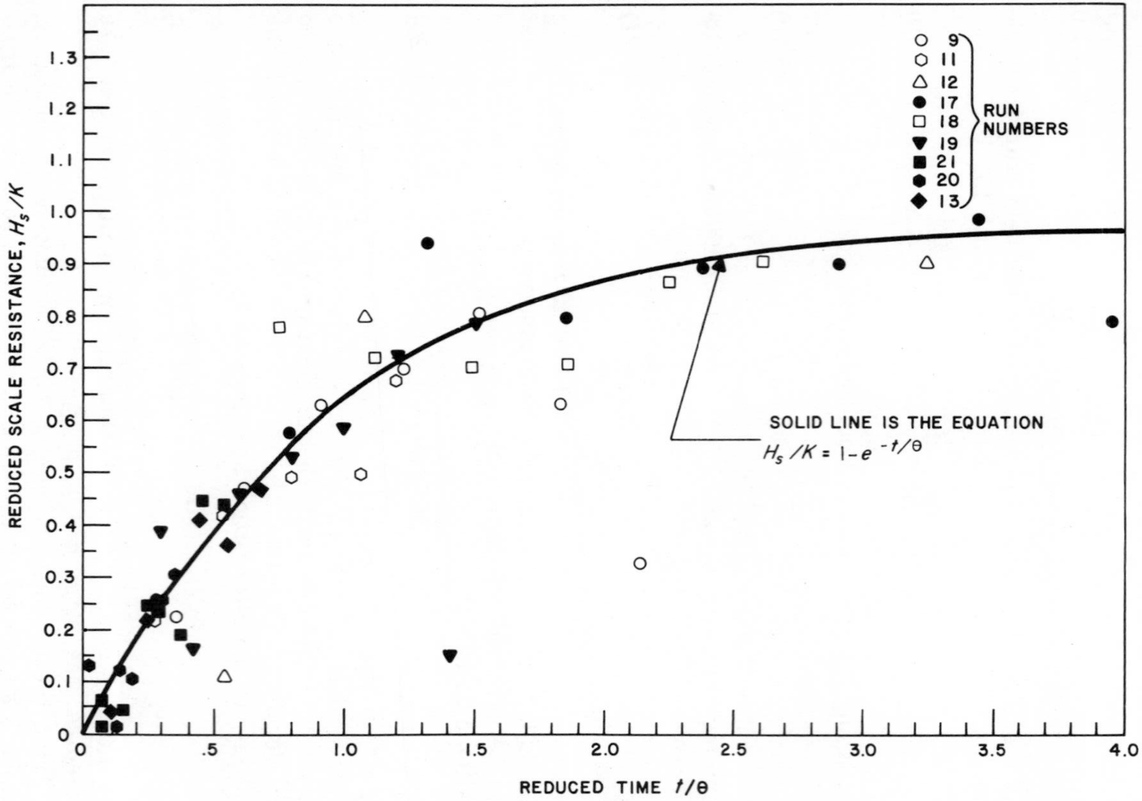

Figure 5.16 Correlation of heat transfer resistance due to silica scale deposition on a cooled surface with Equation 5.24.

from 1 to 3 times that shown in Table 5.11, over pH range from 6.8 to 7.8, at Reynolds numbers from 7000 to 41,000, and log mean temperature differences varying from 38 to 90°F all correlate well with Equation 5.24 as shown in Figure 5.16.

Summary of Mechanisms

Since constant heat flux is not a typical process condition, the rate of deposition of material from a geothermal brine will occur at an exponentially decreasing rate for a cooled or heated surface assuming all process conditions remain constant. For a no-heat flux surface, the deposition rate will be constant. The rate of deposition is proportional to the degree of supersaturation where the proportionality constant may be a rate of reaction, as in the case of a slow reaction such as silica, may be the mass transfer diffusion constant for the case of fast kinetics such as calcite deposition, or may be a combination of the two.

CONTROL OF DEPOSITION

Calcium carbonate deposition can be controlled by controlling pH because increased carbonate concentration caused by a pH increase associated with removal of CO_2 causes deposition. This effect will usually exceed the increase in solubility due to the decreased temperature of the solution. Accordingly, deposition can be decreased by preventing a change in pH of the solution due to flashing and removal of carbon dioxide with attendant pH changes. Alternatively, it may be possible to add an acid to maintain the pH sufficiently low to maintain solubility. Because calcium carbonate deposition is rapid, the problem is a critical one for geothermal brines that have sufficient quantity of calcium and carbonate to form calcium carbonate scale. Because of the large quantity of brine being processed and its relatively low unit value, the use of chemicals to effect pH control may not be economically practical. In general, it has been found that maintaining pressure on a geothermal brine solution prevents precipitation not only of calcite or calcium carbonate, but also of other hydroxide salts as well as certain silicates. Although pure silica deposition is not affected by pH, silica coprecipitation and coaggulation with cations and cationic solutions is affected by pH.

Knowledge of the mechanism of scale deposition means that it may be possible to design a process condition such that deposition will occur in the bulk phase, that is, in the liquid rather than on the walls of the container so that the solids can be collected in a special vessel. For example, if the brine is allowed to flash a sufficient amount so that the solution is greatly supersaturated with respect to calcium carbonate then nucleation may occur in the bulk phase rather than only on the walls of the container, thus forming calcite crystals in the bulk phase. An example of this is the decrease in deposition downstream of a flash separator tank for high degrees of flashing as shown by the upper lines in Table 5.9.

Another example is control of silica deposition. Silica deposition can be controlled by a variety of means because of its relatively slow kinetics. At Otake, for example, a large holding tank was constructed in which the geothermal brine is maintained for a sufficiently long time so that the silica will deposit in the bulk of the solution. This colloidal precipitate is carried through the process equipment in the solution without deposition, thus avoiding pipeline deposition that was a severe problem without the holdup tank system. This method does not always solve the problem (Mahon et al., 1975), because the character of the colloid formed is important in subsequent deposition. Similar effects but on a different time scale would be expected with calcite deposition. Wahl and Baker (1971) have studied agglomeration of titanium dioxide solutions and similarly concluded that the subsequent agglomeration depends on the character and size distribution of the solution. For calcium carbonate precipitation, homogeneous nucleation predominates at high metal ion concentrations but at lower concentrations heterogeneous nucleation predominates (Packter, 1975).

In New Zealand, the DSIR is experimenting with removal of silica by the addition of lime (Rothbaum, 1975). The resulting calcium silicate is filtered and dried to produce an amorphous powder with a bulk density of $0.2\,g/cm^3$. The Broadlands brines produced gels that settled better than the Wairakei gels possibly because of the temperature difference. Low addition rates of lime give silica-rich calcium silicates, while excess lime precipitates stoichiometric calcium silicate.

In general, the approach in Iceland for control of deposition from geothermal brines has been to either maintain the geothermal brine above the temperature at which deposition would occur or to dilute the brine such that the concentration of the salts is maintained below their solubility limit. Early scale deposition problems have not been significant. The silica deposition in the recently developed reservoirs at Namafjall and Hveragerdi, Table 5.2, have shown that silica deposition is a relatively serious problem. The deposits have been removed by wire brushing at 3- to 6-month intervals in the small heat exchangers and after several years of operation in the large distributor pipeline. The addition of 35% cold water before flashing to atmospheric pressure reduced the scaling problem at Hveragerdi. At Namafjall, the proposed solution is to heat fresh water in a tubular heat exchanger and use that as the heat distribution fluid (Thorhallsson et al., 1975) rather than using the brine directly.

In the past, the approach to controlling scale deposition from geothermal brine has been to avoid the problem by either not using wells that show severe scaling problems or by maintaining the temperature above the point of precipitation. In all these cases, however, significant energy is not utilized. Thus considerable advantage can be gained in utilizing geothermal brine if methods can be found for controlling the deposition other than by dilution or temperature. Chemical means of control may not be economic but are possible. Since the maximum value of a 350°F brine at $3/1,000,000 Btu is about $400/1,000,000 lb, it is economically feasible to add about $40 of chemical/1,000,000 lb, which would amount to the addition of 40 ppm of a $1/lb chemical.

Small amounts of impurities in solutions can have a marked effect on the rate of homogeneous nucleation. Heterogeneous nucleation may be retarded by additives that tend to absorb on the growth sites of scale, thereby retarding the growth process. A common example of this effect is found in the addition of leveling agents to electroplating baths. Such substances as glue, gelatin, gum arabic, and so on, are often added to such baths to prevent the formation of nodules and give a leveling effect to the electroplate. It is assumed that these materials attach themselves preferentially to growth sites thereby preventing their further growth. Similar effects might be expected to occur in the growth of scale by deposition from solution.

The best approach to scale deposition is to alter process conditions, based on an understanding of the mechanism and kinetics of the deposition, to control the formation of the deposit in a desired location where mechanical removal can be easily achieved. The relationship between process conditions and deposition is exemplified by the detailed example of deposition in the experiments at East Mesa well 6-1 described in an earlier section. Control of silica deposition as described previously exemplifies the use of process conditions for control. The minimization of scale might also be accomplished by reducing the rate of nucleation on solid surfaces thereby preventing the beginning of the formation of scale. Alternatively, accelerating the rate of nucleation in the bulk of the solution will tend to deplete the solution of scale-forming constituents and thereby decrease the rate of growth of scale on heterogeneous nuclei, that is, existing scale. Also, increasing the number of nuclei that are formed homogeneously in the solution will result in the formation of more and smaller particles, which should have a less detrimental effect on downstream equipment.

SUMMARY

The deposits of insoluble salts from aqueous solutions may occur in the bulk of the solution as a granular or flocculent precipitate or as a scale on an already-formed substrate, either on suspended precipitate in the brine or on a solid surface. Scale deposition from geothermal brines depends on the brine composition, temperature, pressure, and process conditions. Relationships between brine chemistry, process conditions, and scale deposition are exemplified by experimental results. Silica and silicate deposits depend on the content of minor cationic constituents. A synthesized brine deposits silica according to an exponential model. Brine from East Mesa well 6-1 deposited calcite at a constant rate that depended on flashing of the brine and separation of the vapors, but was independent of heat flux. The chemical and physical nature of the deposits show that the deposition is quite sensitive to process conditions.

In most cases, the surface will not affect the deposition rate so that the rate equation will simply be a linear one with time in which the deposition rate is given by an equation of the form:

$$R_D = k(C_B - C_S^*) \tag{5.8}$$

One case in which surface effects, other than nucleation, will cause a change in deposition rate as scale deposits is that in which there is a heat flux and the deposition rate depends on surface temperature. In this case, the deposition will be initially of exponential form. In the past, control of deposition has been limited. However, the need for deposition control for exploiting geothermal hydrothermal reservoirs has been recognized and work is beginning.

Kinetics, pH, thermal shock, and chemical composition can be individually or collectively used to control scale.

NOMENCLATURE

a_i an arbitrary constant
c_p heat capacity, Btu/lb°F
C_B concentration in the bulk of the solution
C_S^* effective concentration for use in the deposition equation, approximately the concentration in equilibrium with the solid at a surface
D molecular diffusivity, ft^2/hr
h heat transfer coefficient, Btu/hr ft^2 °F
H_i heat transfer resistance for material i, Figure 5.15, hr ft^2 °F/Btu
k rate constant for mass transfer or chemical reaction or some combination thereof
k_c rate constant for precipitation of silica
k_i rate constant for the ith chemical reaction;
k_D mass transfer diffusion rate constant, lb mol/hr ft^2
k_T thermal conductivity, Btu/hr ft °F
K constant in Equations 5.21 to 5.24
M_D molecular weight of the diffusing salt
q heat flux, Btu/hr ft^2
R_D rate of deposition at a surface, lb mol/ft^2 hr
R_S rate of dissolution from a surface, lb mol/ft^2 hr
S scale thickness
t time, hr
t_r reduced time
T temperature, °R
w_B weight percent concentration in bulk of solution
w_B^* weight percent equivalent of C_S^*

Greek Symbols

μ viscosity, lb/ft hr
ρ density, lb/ft^3
θ time constant, hr

Subscripts

B bulk portion of brine solution
C coolant

D deposition or diffusion
O at time zero
S surface
M monomeric silica
P polymerized silica

REFERENCES

Badger, W. L., and Associates, Inc., "Critical Review of Literature on Formation and Prevention of Scale," OSW R&D Report No. 25, July 1959.

Beal, S. K., "Deposition of Particles in Turbulent Flow on Channel or Pipe Walls," *Nucl. Sci. Eng.*, **40**, 1–11 (1970).

Cuellar, G., "Behavior of Silica in the Geothermal Waste Waters," Second U.N. Symposium on the Development of Geothermal Resources, San Francisco, May 20–29, 1975.

Friedlander, S. K., and H. F. Johnstone, "Deposition of Suspended Particles from Turbulent Gas Streams," *Ind. Eng. Chem.*, **49**, 1151–1156 (1957).

Hasson, D., et al., "Mechanism of Calcium Carbonate Scale Deposition on Heat Transfer Surfaces," *Ind. Eng. Chem. Fund.*, **7**, 59–65 (1968).

Hasson, D., and J. Zahavi, "Mechanism of Calcium Sulfate Scale Deposition on Heat-Transfer Surfaces," *Ind. Eng. Chem. Fund.*, **9**, 1–10 (1970).

Hatch, L. P., and G. G. Weth, "Scale Control in High Temperature Distillation Utilizing Fluidized Bed Heat Exchangers," OSW R&D Report No. 571, July 1970.

Hermannsson, S., "Corrosion of Metals and the Forming of a Protective Coating on the Inside of Pipes Carrying Thermal Waters used by the Reykjavik District Heating Service," *Geothermics*, Special Issue 2, **2** (2), 1608 (1970).

Kryukov, P. A., and Larinov, E. G., "Physico-Chemical Sampling of High Temperature Wells in Connection with Their Encrustation by Calcium Carbonate," *Geothermics*, Special Issue 2, **2** (1), 1624 (1970).

Langelier, W. F., D. H. Caldwell, and W. B. Lawrence, "Scale Control in Seawater Distillation Equipment," *Ind. Eng. Chem.*, **42**, 126 (January 1950).

Lin, C. S., et al., "Mass Transfer Between Solid Wall and Fluid Streams," *Ind. Eng. Chem.*, **45** (3), 636–640 (1953).

Mahon, W. A. J., et al., "Silica Deposition Discharged from Geothermal Wells," Second U.N. Symposium on the Development and Utilization of Geothermal Resources, San Francisco, May 1975.

Owen, L. B., "Precipitation of Amorphous Silica from High Temperature Hypersaline Geothermal Brines," Lawrence Livermore Laboratory, UCRL-51866, June 1975.

Ozawa, T., and Y. Fujii, "A Phenomenon of Scaling in Production Wells and the Geothermal Power Plant in the Matsukawa Area," *Geothermics*, Special Issue 2, **2** (2), 1614 (1970).

Packter, A., "The Precipitation of Calcium Carbonate Powders from Aqueous Solution with Slow Development of Supersaturation," *Krist. Tech.*, **10**, 111–121 (1975).

Rothbaum, H. P. and B. H. Anderton, "Removal of Silica and Arsenic from Geothermal Discharge Waters by Precipitation of Useful Calcium Silicates," Second United National Symposium on the Development and use of Geothermal Resources, San Francisco, May 1975.

Skinner, B. J., D. E. White, and H. J. Rose, "Sulfides Associated with the Salton Sea Geothermal Brine," *Econ. Geol.*, **62**, 316 (1967).

Standiford, F. C., and J. R. Sinek, "Stop Scale in Seawater Evaporators," *Chem. Eng. Progr.*, **57**, 58 (1961).

Taborek, J., et al., "Fouling: The Major Unresolved Problem in Heat Transfer," *Chem. Eng. Progr.*, **68** (2), 59–67 (1972).

Taborek, J. T., et al., "Predictive Methods for Fouling Behavior et al.," *Chem. Eng. Progr.*, **68**, 69–78 (1972).

Thorhallson, S., et al., "Rapid Scaling of Silica in two District Heating Systems," Second U.N. Symposium on the Development and Use of Geothermal Resources, San Francisco, May 1975.

Van Hook, A., *Crystallization*, Reinhold, New York, 1961, pp. 1148.

Wahl, E. F., and C. G. Baker, "The Kinetics of Titanium Dioxide Agglomeration in an Agitated Liquid Suspension," *Can. J. Chem. Eng.*, **49**, 742–746 (1971).

Wahl, E. F., G. G. Pangborn, and H. A. McCurdy, *J. Paint Tech.*, **44**, 98 (1972).

Wahl, E. F., I. K. Yen, and W. J. Bartel, "Silicate Scale Control in Geothermal Brines," Contract No. 14–30–3041, Office of Saline Water, U.S. Department of the Interior, Washington, D.C., September 1, 1974.

Wells, A. F., *Crystal Chemistry*, third ed., Oxford Univ. Press, London, 1962, Chapter 21.

Yanagase, T., et al., "The Properties of Scales and Methods to Prevent Them," *Geothermics*, Special Issue 2, **2** (2), 1620 (1970).

CHAPTER 6

Mechanical or Electrical Work from a Thermal Source

Most electrical and mechanical power is generated from thermal sources. Examples are coal, oil, gasoline and nuclear fuels, all of which first produce a high temperature fluid. Exceptions to this are hydrogenerating systems such as are powered by tides or elevated water sources like rivers. Geothermally heated waters in the crust of the earth are another example of a thermal source. Resources that produce, or are in the form of, a high temperature fluid may be converted into mechanical work and then electricity as shown in Figure 6.1. Electrical energy is generally produced by a generator. The generator in turn is powered from an expansion machine that is producing mechanical work in the form of a rotating shaft. Examples of expansion machines are steam turbines and piston-driven engines. The hot fluid must be at a higher pressure, such as P_1 in Figure 6.1, than the exhaust pressure of the expansion machine, if the thermal energy of the fluid is to be recovered by expanding it through the machine. This elevated pressure of the hot fluid can be obtained in a number of ways. One method, as in the combustion of coal or oil in conventional power plants, is to transfer the heat from the combusting fluid at atmospheric pressure through a heat exchange surface to a

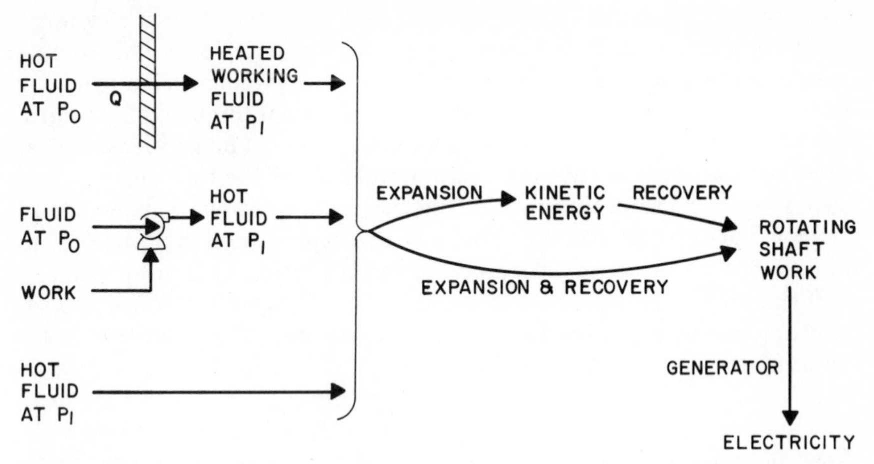

Figure 6.1 Methods for converting thermal fluids to mechanical work or electricity.

working fluid at an elevated pressure. Another method is to pump the combustible material, such as gasoline, to a higher pressure and then burn it to produce a hot fluid at a sufficiently high pressure that it can be expanded directly in the expansion machine. A third means of expansion is available if the fluids, as in the case of geothermal brines, are already available at a sufficiently high pressure so that they can be directly expanded through the machine without further pumping. In any case, the conversion of thermal energy into mechanical and electrical work is accomplished by expanding a hot fluid to produce mechanical work.

The expansion of a hot fluid to produce mechanical work is accomplished by converting its thermal energy into kinetic energy and then recovering this kinetic energy in the form of mechanical work by reducing the discharge velocity of the fluid to a minimum. In some cases, the mechanical work is produced directly in mechanical form during the expansion process. A good example of this is the internal combustion engine in which the high temperature fluid is expanded to move the piston and so produce mechanical work directly.

FIRST LAW OF THERMODYNAMICS

The first law of thermodynamics is the conservation of energy. This means that the sum of all of the energies including thermal, mechanical, electrical, gravitational, and any other forms of energy must be a constant in the absence of nuclear reactions. This simple statement is a powerful tool for analyzing energy conversion systems. Applied to the simple but general system shown in Figure 6.2, the first law of thermodynamics, that is, the conservation of energy, states that the total energy flow into the system must equal the total energy flow out of the system providing there is no accumulation or depletion of energy within the system. The process represented by the box in Figure 6.2 could be any very complicated process or a simple one.

If the process is operating at steady state, that is the process plant is running steadily and not fluctuating and its operating conditions such as temperatures, pressures, flow rates, and so on, are constant, then the sum of all of the energies flowing into the system must equal those flowing out. Taking as a basis some unit of time, such as 1 hr, then the flows of mass and energy become quantities of mass and energy. These energies are then conveniently classified as thermal energies represented by the symbol Q, by mechanical energies such as shaft work or electricity represented by the symbol W^*, and by quantities of mass represented by the symbol m with their associated energy. The mass flow streams will commonly contain energy in the form of chemical, gravitational, and/or kinetic energy. They will always contain energy in the form of internal energy and the pressure volume work that must be done against the surroundings; that is, if the pressure and/or

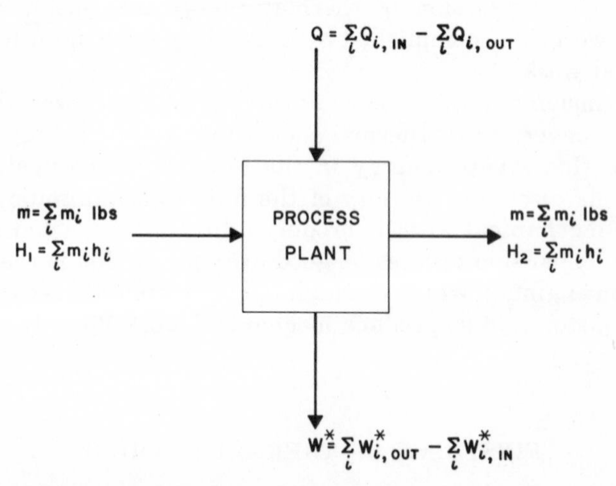

Figure 6.2 Energy transport into and out of a steady-state process plant. A definition of the terms used are discussed in the text.

volume of the fluid entering the process is different than that leaving the process, then an amount of work must have been done on the surroundings equal to the pressure of the fluid times the volume transported:

$$W^* = \frac{PV}{J} \tag{6.1}$$

The combination of the internal energy plus the pressure volume work is called enthalpy and is given the symbol H:

$$H = E + \frac{PV}{J} \tag{6.2}$$

The thermodynamic properties per unit mass of a fluid are commonly given lower case symbols, whereas the total content of a stream is given capital letters. Thus

$$H = \sum_i m_i h_i \tag{6.3}$$

Conventionally, the work terms are defined as being positive for outflowing work, and the thermal terms are defined as being positive for inflowing heat.*

Applying the first law of thermodynamics using this terminology to the steady-state system shown in Figure 6.2 results in

$$H_2 - H_1 = Q - W^* \tag{6.4}$$

Of course the total mass flowing into the process must equal the total mass flowing out. In energy conversion systems, we are generally dealing with only one fluid stream entering the process and one leaving. In this case, Equation 6.4 can be rewritten as

$$m\Delta h = Q - W^* \tag{6.5}$$

where Δh is the enthalpy change of the fluid as a result of passing through the process plant. If the fluid contains other energy forms such as kinetic energy or potential energy, these must be added to the left-hand side of Equations 6.4 or 6.5.

*Because of the repeated use of exhaust heat flowing out of the systems that have no heat inflow, the exhaust heat from the system per pound of geothermal brine flowing through it is given in following chapters with the symbol Q_{EX} and is defined to be positive for outflowing heat. Q_{th} is also defined as positive for outflowing heat.

AVAILABLE WORK

A geothermal energy source delivers its stored thermal energy as a hot fluid at some temperature and pressure to a process system. The maximum energy will have been extracted from this fluid when it is reduced to atmospheric pressure and the temperature of the surroundings. Such a process is represented schematically in Figure 6.3. The maximum mechanical work will have been extracted from this brine if the energy is recovered and converted into work in a reversible fashion. "By reversibility" means that all of the equipment operates in a frictionless manner and that all heat is transferred reversibly. Reversible heat transfer requires that all heat transport across boundaries between fluids occur at infinitesimally small temperature differences.

The work obtained from a process such as is shown in Figure 6.4 and which utilizes a hot fluid is calculated by applying the first law of thermodynamics, Equation 6.5, to the entire process enclosed by the dashed lines. The result is

$$m\Delta h = Q_0 - W^* \tag{6.6}$$

Note that this equation applies no matter what the process within the boundaries of Figure 6.4 is and whether or not it is reversible.

One of many possible examples of a reversible scheme for converting geothermal fluid at some pressure P and temperature T to obtain the maximum work consists of the process steps shown in Figure 6.4. The first step in the process is to reversibly process the fluid from its higher pressure to the pressure P_0 of the surroundings adiabatically, that is, with no heat flow into or out of that process step. The next step in the process is to reduce the temperature to that of the surroundings

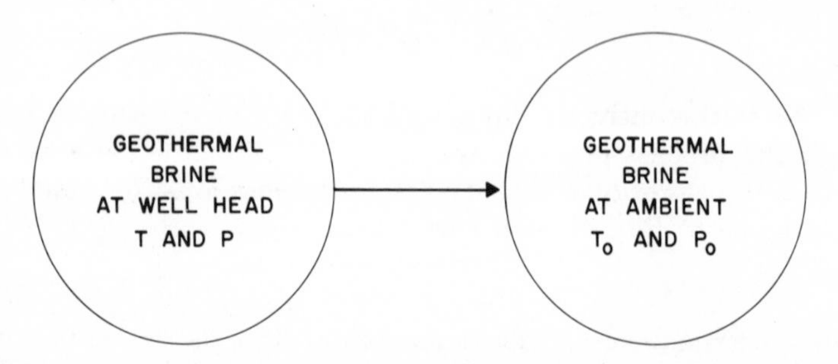

Figure 6.3 A hot fluid contains thermal energy that can be utilized by extracting it and leaving the fluid at ambient temperature and pressure conditions.

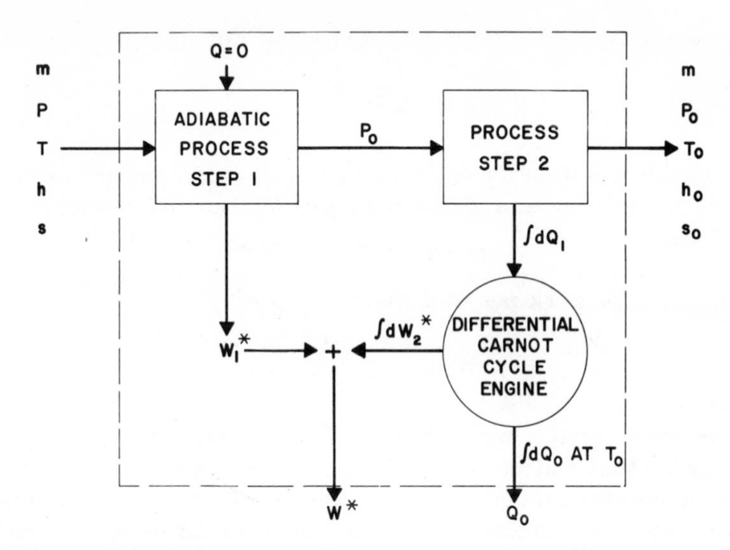

Figure 6.4 A reversible process for obtaining the maximum amount of work from a geothermal brine delivered to the wellhead at the pressure P and temperature T.

reversibly. This can be done by transferring the heat Q_1 through a reversible work machine and then rejecting the heat from that machine to the surroundings at a temperature equal to the temperature of the surroundings.

The entropy change of a substance undergoing a reversible process is defined as the heat flow into the process divided by the absolute temperature:

$$dS = \frac{dQ_R}{T} \tag{6.7}$$

Application of the second law of thermodynamics, which can be stated as the fact that heat will flow only in the direction of decreasing temperature, shows that entropy is a state property of the fluid. Since the process of Figure 6.4 is a reversible one, the total heat flow out is the reversible heat flow:

$$\int dQ_R = \int dQ_0 \tag{6.8}$$

Since the reversible heat flow from the process is being delivered at the constant temperature of the surroundings, the entropy change for the process can be obtained by integrating Equation 6.7 to give

$$\Delta S = \frac{Q_0}{T_0} \tag{6.9}$$

Solving this equation for Q_0, making the change from total values to values per unit mass, and inserting in Equation 6.6 results in

$$m\Delta h = mT_0\Delta s - W^* \tag{6.10}$$

Solving for the work per unit mass,*

$$W = \frac{W^*}{m} = T_0 s - \Delta h \tag{6.11}$$

This is the reversible work obtained from a unit mass of geothermal fluid, or for that matter any fluid when it is processed reversibly from some temperature and pressure to ambient conditions. This applies no matter what the process is within the boundaries of Figure 6.4, providing that it is a reversible one. Since the process is reversible, this is the maximum work that can be obtained from the fluid and is termed the available work.

The available work of water as a function of temperature is plotted in Figure 6.5. Since dilute geothermal brines behave thermodynamically as water, Figure 6.5 gives the theoretical maximum work that can be produced in a process unit of any sort from a dilute geothermal brine at specified well temperature and pressure conditions. For this case, the available work is calculated using Equation 6.11 and the entropy and enthalpy of water at ambient conditions:

$$W = (h - h_0) - T_0(s - s_0) \tag{6.12}$$

In this equation, the sign of the terms within the brackets has been reversed from prior equations for convenience of future use and because it more correctly portrays the source of the work. Because a fluid such as a geothermal brine enters a process at a temperature above the exhaust temperature, the entering enthalpy and entropy are greater than the exhaust enthalpy and entropy. Thus the terms in Equation 6.12 are positive as written. The term $h - h_0$ represents the total thermal energy, and the term $T_0(s - s_0)$ represents the thermal energy not converted to work in a reversible process.

Figure 6.5 shows that the available work for water with a wellhead

*Since geothermal utilization processes usually have one mass flow stream, namely the brine, the remainder of the text generally deals with the work produced per pound of brine processed W.

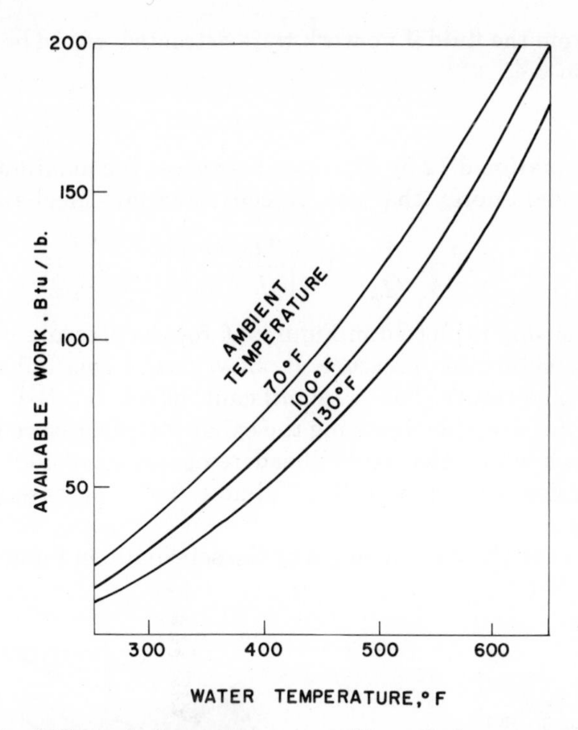

Figure 6.5 Available work, the maximum reversible work, for water at its saturation pressure and temperature discharging to ambient conditions of 1-atm pressure and various temperatures as shown.

temperature of 500°F, discharging at an ambient temperature of 100°F, is 100 Btu/lb compared with 25 Btu/lb for a 300°F wellhead temperature. At 400°F, the available work doubles for an increase of 100°F. This illustrates the significance of the temperature of the geothermal reservoir and the importance of a small increase in wellhead temperature.

To evaluate the available work for a given reservoir from the reservoir conditions, it is necessary to correct the pressure and temperature for the potential energy difference between the reservoir and the surface. If the flow is reversible from the reservoir to the surface, then the available work can be calculated from the reservoir conditions less the work necessary to raise the fluid against the hydrostatic head.

It is interesting to compare the available work as given in Figure 6.5 with the thermal energy content of the water. The thermal energy content of the water, that is, the total thermal energy that could be obtained from the fluid if no work were extracted, would be, according to Equation 6.4,

$$Q_{th} = h - h_0 \tag{6.13}$$

Dividing Equation 6.12 by Equation 6.13 gives the maximum fraction of the thermal energy that can be converted into mechanical work

$$\frac{W}{Q_{th}} = 1 - \left[\frac{T(s - s_0)}{(h - h_0)}\right] \tag{6.14}$$

This relationship is plotted in Figure 6.6 for the pressure of the water equal to its saturation pressure. It shows that a small change in the exhaust temperature has a significant effect on the conversion efficiency. For example, lowering the exhaust temperature from 130°F to 100°F for a water, that is, wellhead, temperature of 400°F increases the conversion efficiency 12% or almost 0.5% increase per degree Fahrenheit.

The more general case depicted by the schematic in Figure 6.7 better

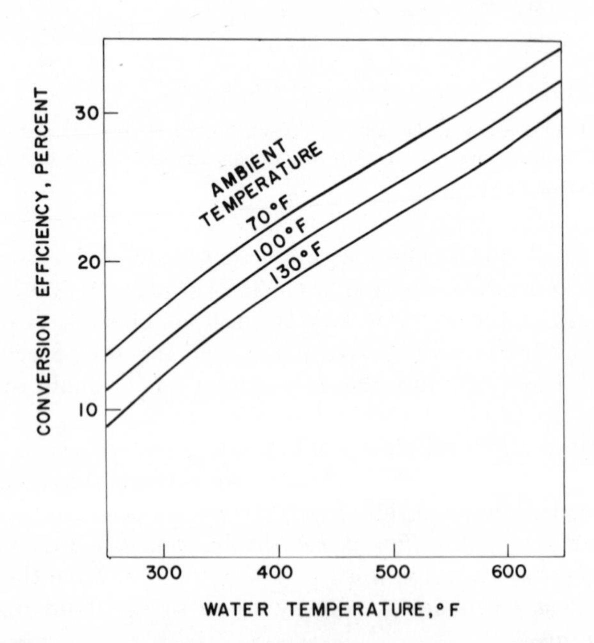

Figure 6.6 The conversion efficiency of the thermal energy content of water above ambient temperature to work as a function of water temperature and at a pressure equal to its saturation pressure. A reversible process discharging the water at atmospheric pressure and ambient temperature is assumed.

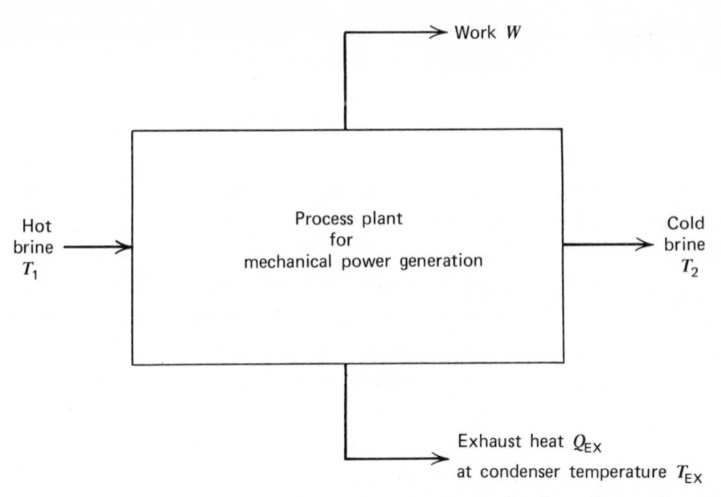

Figure 6.7 General process for mechanical power generation from thermal energy of a geothermal brine. The nomenclature is defined as shown for the inlet and outlet streams and properties.

represents a geothermal energy utilization process, because both the spent-brine discharge temperature T_2 and the heat exhaust temperature T_{EX} are above the ambient temperature T_0. Applying the same logic as used to derive Equation 6.12, the reversible work W per pound of brine entering the process at temperature T_1 is given by

$$W = (h_1 - h_2) - T_{EX}(s_1 - s_2) \tag{6.15}$$

The importance of variation in the temperatures T_1, T_2, and T_{EX} on the work output is shown by graphical presentations in Chapter 8 or an analytic expression for the fractional change in work output per unit change in the jth temperature, namely, $(dW/dT_j)/W$. This function can be obtained by evaluating the right-hand side of Equation 6.15 assuming a constant heat capacity c of the brine and dividing the differential of the result by itself. Since

$$h_1 - h_2 = c(T_1 - T_2) \tag{6.16}$$

and

$$(s_1 - s_2) = \int_{T_1}^{T_2} \frac{c}{T}\, dT$$
$$= c \ln (T_1/T_2) \tag{6.17}$$

Equation 6.15 becomes

$$W = c[T_1 - T_2 - T_{EX} \ln (T_1/T_2)] \tag{6.18}$$

Approximating $\ln x$ by the first term $(x - 1)/x$ of the power series expansion of $\ln x$, Equation 6.18 becomes

$$W \simeq c(T_1 - T_2)\left(1 - \frac{T_{EX}}{T_1}\right) \tag{6.19}$$

By differentiating this expression with respect to the appropriate temperature and dividing the result by the expression for work, the results in Table 6.1 are obtained. For a typical case of 400°F wellhead water, a brine discharge temperature of 180°F, and a condenser temperature of 130°F, the last column of Table 6.1 shows that a unit change in temperature T_1 delivered to the process is 7 times as effective as a unit change in condenser temperature T_{EX}, which in turn is 3 times as effective as a unit change in temperature T_2 of the spent brine. These relationships apply whether or not the process is 100% efficient, providing the efficiency does not change with the temperature under consideration.

Table 6.1 Values of $(dW/dT_j)/W$, the Fractional Variation in Work Output with Temperature for Various Temperature T_j

T_j	$\frac{1}{W}\frac{dW}{dT_j}$ General Approximation	$\frac{1}{W}\frac{dW}{dT_j}$ Typical Case[a]
T_{EX}	$-\frac{1}{T_1 - T_{EX}}$	$-\frac{1}{T_1 - T_{EX}}$
T_2	$\frac{1}{T_1 - T_{EX}}\left(\frac{T_1}{T_2}\frac{T_{EX} - T_2}{T_1 - T_2}\right)$	$-\frac{1}{3}\frac{1}{T_1 - T_{EX}}$
T_1	$\frac{1}{T_1 - T_{EX}}\left(\frac{T_1 + T_{EX}}{T_1 - T_2}\right)$	$7\frac{1}{T_1 - T_{EX}}$

[a] $T_1 = (400 + 460°)$F; $T_2 = (180 + 460)°R$; $T_{EX} = (130 + 460)°$R.

SUMMARY

The conservation of energy, the first law of thermodynamics, applied to a steady-state process plant into and out of which m pounds of material are flowing without kinetic or potential energy effects states that the change in enthalpy between the entering and leaving material streams is equal to the total heat input to the process minus the total net work output:

$$m \Delta h = Q - W^* \tag{6.5}$$

The maximum work will be obtained from a process when it is operating reversibly, that is, when all friction losses are negligible and temperature and heat transfers are being carried out with infinitesimally small temperature differences. This reversible work output, determined by applying the first law of thermodynamics and the definition of entropy, is equal to the enthalpy of the hot brine above its enthalpy at ambient temperature less the product of the absolute temperature and the hot brine entropy above its entropy at ambient conditions:

$$W = (h - h_0) - T_0(s - s_0) \tag{6.12}$$

This is the maximum work output that can be obtained from one pound of geothermal fluid. Furthermore, this work output is the work that is produced from any processing plant no matter what its configuration nor what equipment is used providing that it operates reversibly.

NOMENCLATURE

E internal energy, Btu
h specific enthalpy, Btu/lb
H enthalpy, Btu
J conversion factor, 778 ft lb-ft/Btu
m mass, lb
P pressure lb-f/ft^2
Q heat, flowing into process, Btu; positive for inflowing heat
Q_{th} heat produced per pound of brine processed, Btu/lb; positive for outflowing heat
s specific entropy Btu/lb °R
S entropy, Btu/°R
T absolute temperature, °R
V volume, ft^3
W^* work, Btu
W specific work, Btu/lb

Subscripts

i ith component
M maximum

0 ambient conditions
R reversible

REFERENCE

Weber, H. C., and H. P. Meissner, *Thermodynamics for Chemical Engineers*, second ed., Wiley, New York, 1959.

CHAPTER 7

Expansion Machines

Expansion machines are used for converting the thermal energy of a hot pressurized fluid into rotating mechanical shaft work. Over the past 100 yr, a variety of mechanical devices have been invented, developed, and used for this purpose. Prominent examples are the reciprocating piston steam engine and steam or gas turbines. In general, all of the types can be classified either as a device in which the mechanical work is recovered as the fluid expands or as a device that first converts the thermal energy into kinetic energy and then recovers the kinetic energy in the form of mechanical work as shown schematically in Figure 6.1.

For these machines to operate at best efficiency, the heat losses from the machine and the fluid friction losses during the expansion process should be kept to a minimum. Modern expansion machines with their high throughputs operate essentially as adiabatic processes. Applying the first law of thermodynamics, Equation 6.4, to the adiabatic expansion process represented schematically in Figure 7.1 gives*

$$W = h_1 - h_2 \tag{7.1}$$

*Since geothermal brine processing plants as discussed henceforth have only one inflowing and outflowing process stream, the work W will be defined as the work output per pound of brine flowing through the process.

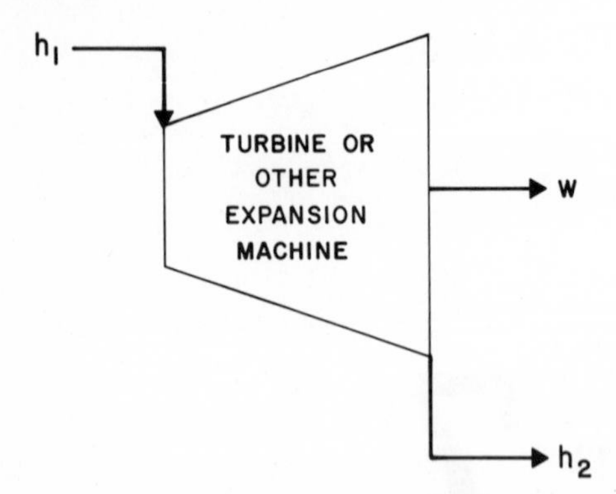

Figure 7.1 Schematic of energy flow for an adiabatic expansion machine such as a turbine.

If the expansion of the fluid in the machine is reversible, then, according to Equation 6.7, the entropy change for the process will be zero because the heat effects are zero in an adiabatic process. Thus if enthalpy and entropy data are available for the working fluid as a function of pressure and temperature, then it is a simple matter to evaluate the adiabatic reversible work by applying Equation 7.1 according to

$$W_R = h_1 - h_{2R} \qquad \text{for} \qquad \Delta S = 0 \tag{7.2}$$

This is shown schematically by the dashed line on the enthalpy–entropy diagram Figure 7.2. In any real expansion process, the entropy change will be greater than zero due to the irreversible losses, particularly the fluid friction losses. Thus the actual entropy change for a given pressure ratio will be given by the solid line in Figure 7.2 rather than the dashed line for the reversible case. The efficiency of the expansion machine is defined as the ratio of the work output to the reversible work output and for an adiabatic process is given by

$$e = \frac{(h_1 - h_{2A})}{(h_1 - h_{2R})} \tag{7.3}$$

Thus the work can be calculated from the reversible enthalpy change and the machine efficiency by the equation

$$W_A = e(h_1 - h_{2R}) \tag{7.4}$$

In the absence of enthalpy–entropy–pressure–temperature data, the reversible expansion work can be estimated by calculating the work done by a gas following a reversible expansion path. For example, the

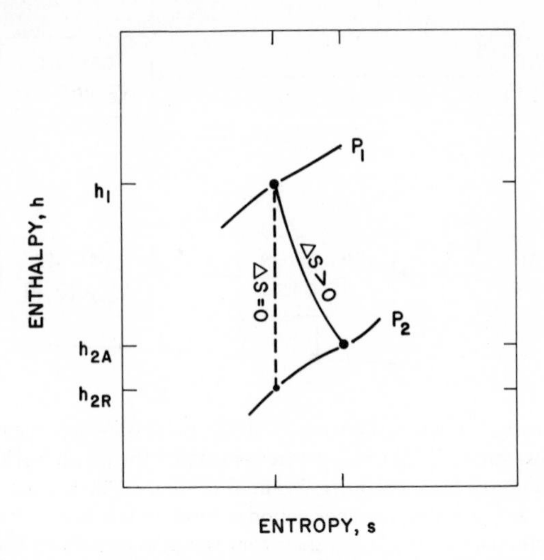

Figure 7.2 Enthalpy–entropy diagram for the expansion of a hot fluid from a pressure P_1 to a pressure P_2. The dashed line represents a reversible process and the solid line a real process.

reversible work of a unit mass of gas expanding from a pressure P to a pressure P_1 in a piston cylinder arrangement to produce the work output W_B is shown schematically in Figure 7.3. The work obtained from the cylinder for the expansion of a unit mass of gas is given by the pressure on the cylinder times its area times the distance through which the piston moves, and since the area times the distance moved is equal to the volume change of the gas, the work output is given by

$$W_B = \int \frac{Pdv}{J} \tag{7.5}$$

For a flow process as shown in the lower portion of Figure 7.3, it is necessary to expel the gas at a pressure P_2 from the apparatus that requires an amount of work equal to the pressure times the volume. In addition, an amount of work will be received due to the pressure of the incoming gas. Thus the reversible work for a flow process is given by

$$W_R = W_B - \frac{(P_2 v_2 - P_1 v_1)}{J} \tag{7.6}$$

Since

$$Pdv = d(Pv) - vdP \tag{7.7}$$

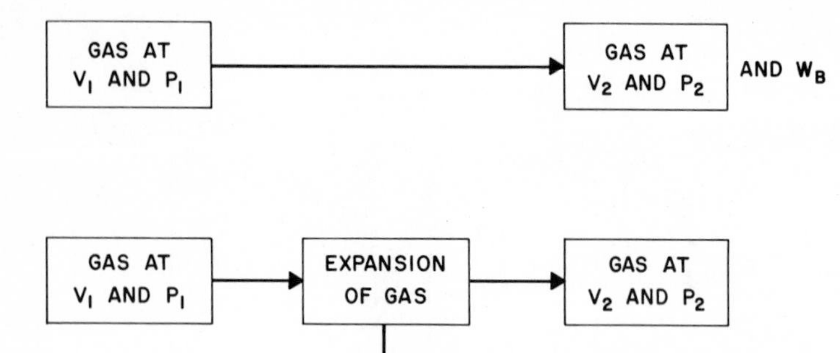

Figure 7.3 Processes for obtaining work from the reversible expansion of a gas from a pressure P_1 to a pressure P_2. The upper line represents the batch work that is obtained by expanding 1 lb of gas from the initial to final pressure. The bottom line of the figure represents the work that is obtained per pound of fluid for the steady-state operation of a device that continuously expands the gas from the pressure P_1 to the pressure P_2.

then Equation 7.5 becomes

$$W_B = \frac{[\Delta(Pv) - \int v dP]}{J} \tag{7.8}$$

Inserting this expression for W_B in Equation 7.6, the expression for the mechanical work delivered by the continuous reversible expansion of a fluid is

$$W_R = -\int \frac{v dP}{J} \tag{7.9}$$

To use this equation, the relationship between the volume and pressure of the gas is needed. Experimental studies have shown

$$Pv^n = a \tag{7.10}$$

for the expansion of a gas in turbines, pumps, and compressors. Inserting this in Equation 7.9 results in

$$W_R = \left[\frac{nzRT}{M(n-1)}\right]\left[1 - \left(\frac{P_2}{P_1}\right)^{(n-1)/n}\right] \tag{7.11}$$

For an isentropic, that is, reversible adiabatic, process, the constant n is equal to the ratio of the specific heats:

$$n = \frac{c_p}{c_v} \tag{7.12}$$

For the case of a polytropic expansion as defined by Equation 7.10, the value of n varies from less than to greater than the specific heat ratio given in Equation 7.12. The value of n can be determined (Schultz, 1962) by

$$n = \frac{1}{[b_3 - b_1(1 + b_2)]} \tag{7.13}$$

where

$$b_1 = \frac{zR(e_t + b_2)}{c_p} \tag{7.14a}$$

$$b_2 = \left(\frac{T}{V}\right)\left(\frac{\delta V}{\delta T}\right)_p - 1 \tag{7.14b}$$

$$b_3 = -\left(\frac{P}{V}\right)\left(\frac{\delta V}{\delta P}\right)_T \tag{7.14c}$$

Using these equations, n is evaluated at some average value of the parameters in the Equations 7.13 through 7.14*c*. Then the actual work is calculated by multiplying the reversible work obtained from Equation 7.11 for a polytropic expansion by the machine efficiency e_t to give

$$W_A = e_t W_R \tag{7.15}$$

TURBINES

Turbines are the most widely used device for expanding hot fluids to generate large quantities of power. For very small power plants, internal combustion devices, such as the gasoline engine or diesel engine, are most useful for producing power. Even there, however, the small gas turbine systems are beginning to compete with reciprocating engines. A turbine has a much higher mass flow capacity than does a piston engine of similar dimensions. Because the capital cost of an expansion machine is a significant portion of the equipment cost of a power plant, the turbine is the only practical method for producing large quantities of power.

Because of the high mass throughputs to relatively small surface area, the heat losses from a turbine are negligible and the turbine expansion process is essentially adiabatic. In addition, if properly designed, the kinetic energy effects and other energy effects of the ingoing and outgoing fluid streams are negligible compared to the work of the expansion process. If enthalpy and entropy data are available for the working fluid, then the reversible work output from a turbine can be calculated using Equation 7.2. If not, then the work can be calculated for the reversible polytropic expansion using Equations 7.10 through 7.14. The actual work output is then obtained by multiplying the reversible work output by the turbine efficiency according to Equation 7.4 or 7.15.

The efficiency of a turbine depends on a number of variables that can be described in terms of the mechanism by which a turbine converts fluid energy into rotational mechanical work. A turbine consists of a number of stages, each stage consisting of a stator, sometimes called nozzles, followed by a rotor, sometimes called buckets or turbine blades. Hot fluid expands through a stator or nozzle section and converts its thermal energy into kinetic energy. This kinetic energy is then recovered on the following rotors. The optimum recovery of energy is achieved when the velocity of the rotors is equal to one-half the jet velocity of the gas issuing from the stators. In an impulse turbine, the total pressure drop for each stage is taken across the nozzle so that the rotor blades operate at substantially constant pressure. In a reaction turbine, the pressure drop occurs equally across the rotor blades and the stator section. These are hypothetical limiting cases and in actual practice the turbine operates somewhere between the two cases. There is an optimum velocity for the rotor blades as a function of the jet velocity issuing from the stator section that results in maximum thermodynamic efficiency of the machine. The optimum efficiency of a blade may vary from about 70 to 90%. The loss of efficiency of turbines results from friction losses of the fluid in the stationary and rotating sections, from leakage losses around the stages, and from liquid droplets in the gas, if it is not superheated, that will be accelerated at a lower velocity than the gas in the expansion sections. Consequently, the efficiency of the turbine depends on the diameter of the stages, the dimensions of the blades, the number of stages, the conditions of the working fluid, and other design variables. It is not possible to calculate an efficiency as a function of any one set of design parameters. Experience with steam turbines shows that the turbine efficiency varies between 60 and 85% depending on its capacity and on the inlet pressure of the fluid as shown in Figure 7.4. The relationship between pressure and turbine size is understood if one realizes that the capacity of a given machine will depend on the inlet pressure of the fluid because of the relationship of fluid density, hence total mass flow rate, to the pressure. Figure 7.4 can also be used to estimate the efficiency for fluids other than steam for given pressure

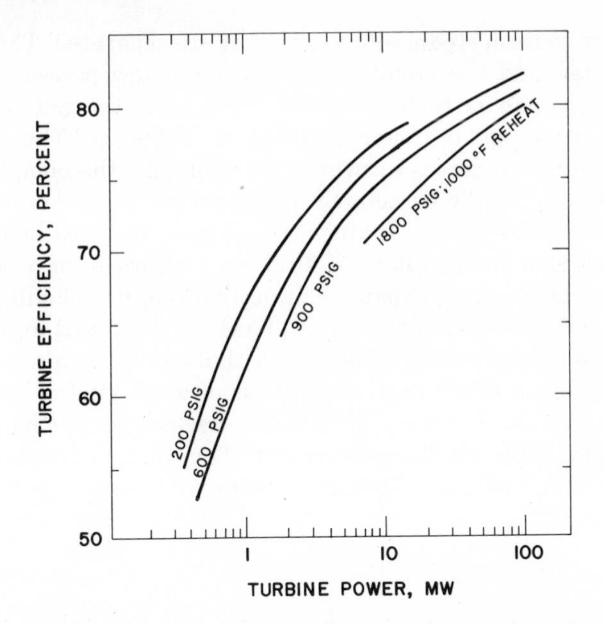

Figure 7.4 Turbine efficiency as a function of the turbine capacity in megawatts for various inlet pressures. Data from Marks and Baumeister (1967) and Perry and Chilton (1973) combined.

conditions. With respect to capacity of the turbine, it should be noted that the capital cost of the turbine per unit power output will increase with decreasing pressure of the entering working fluid. For this reason, the use of a secondary working fluid such as isobutane, at a higher pressure than steam may result in a lower plant cost.

TWO-PHASE EXPANSION MACHINES

Since the thermal energy in hydrothermal reservoirs is generally available as hot water, the use of two-phase expansion machines has been proposed by a number of workers. In the case of the Lardarello and Geysers geothermal fields, steam is produced from the reservoir. This steam can be used directly or after minor cleanup as a working fluid for a low pressure steam turbine as described previously. In the case of most other hydrothermal reservoirs, however, the thermal energy is stored in the form of hot water. To use a conventional gas turbine, it is necessary, therefore, to flash the hot water by dropping its pressure to form steam and water that are separated. The water is then rejected and the steam used to drive a low pressure turbine. Another approach is to transfer the heat from the hot water to a secondary working fluid such as freon or isobutane. The secondary working fluid then drives a turbine. To overcome the complexities of such processes and the associated inefficiencies, a number of workers have proposed two-phase expansion machines. The idea behind these machines is that the geothermal fluid, be it all water or a mixture of steam and water, can be expanded directly through a machine without prior separation to form a pure gas phase. As discussed in a previous chapter, the maximum reversible work that can be obtained from the geothermal brine is the same regardless of what kind of machine or process is used. As is shown by detailed discussions in Chapter 8, all such systems have similar efficiencies. The major advantage of the two-phase machines is a simpler system with lower operating and capital costs.

Two-Phase Nozzle-Driven Reaction Machine

This relatively simple device is described first, since it will also serve as an introduction to the other two-phase machines. If hot water at a sufficiently high pressure is allowed to expand through a nozzle to a lower pressure below the saturation pressure of the hot water, then the water will vaporize during the expansion process. Since the specific volume of water vapor is much less than that of liquid water, only a small fraction of the liquid needs to vaporize to result in the mixture being primarily a gas phase consisting of droplets of water in steam. This is shown schematically in Figure 7.5. The inlet and outlet conditions are shown in this figure. Since the flow rate through the nozzle is very high compared to its surface area, the expansion process will be adiabatic. Furthermore, if the nozzle is stationary, the work output will be zero. Assuming that the inlet velocity is negligible, the first law of thermodynamics, Equation 6.4, for this process becomes

$$\frac{[(1 - X)u^2 + XU^2]}{2gJ} = h_1 - h_{EX} \tag{7.16}$$

If the liquid droplets are accelerated to the same velocity as the gas, then this equation reduces to

$$\frac{U^2}{2gJ} = h_1 - h_{EX} \tag{7.17}$$

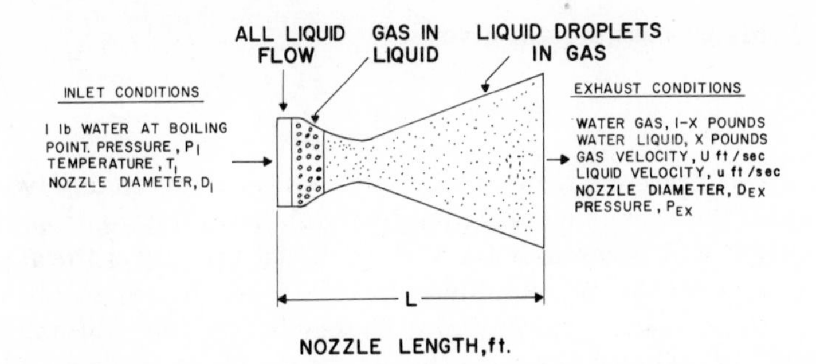

Figure 7.5 Expansion of a liquid fluid such as a geothermal brine from a pressure above its saturation pressure to a pressure much lower than its saturation pressure in a nozzle.

If the expansion process is reversible, which is the case if the fluid friction losses in the expansion are negligible, then the change in entropy for the process will be zero as described by Equation 7.2. If a pair of nozzles are attached to a rotatable shaft as shown in Figure 7.6, then the velocity of the exhaust gases will cause the shaft to turn and power to be transmitted. Writing the energy balance, Equation 6.4,

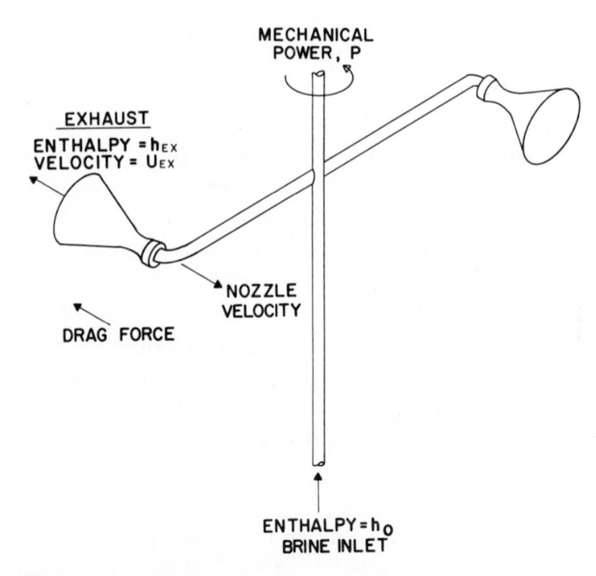

Figure 7.6 Schematic representation of a device for converting the kinetic energy of the exhaust fluid from a nozzle to mechanical work.

around this process system gives

$$W = h_1 - h_{EX} - \frac{U_{EX}^2}{2gJ} \tag{7.18}$$

For a given pressure ratio, the maximum enthalpy change will be achieved if the expansion in the nozzle is adiabatic and isentropic. The work output will be a maximum if the exhaust velocity of the gases is zero relative to the surroundings. In this case, the reversible work output will be equal to the enthalpy change for the liquid between the entrance and exhaust pressures for an isentropic expansion and will be identical in form to that of Equation 7.2. For reasonable temperatures and power outputs, the velocity of the exhaust gases relative to the nozzle will be about 1000 ft/s. To accelerate the liquid droplets to the same velocity as the gas stream would require a very long nozzle. Consequently, for a practical size nozzle, liquid droplets will have a velocity less than that of the gas stream. The result is that there will be fluid friction exerted on the liquid droplets to accelerate them. This will cause an irreversible loss of energy. Furthermore, there will be a gas drag on the nozzles as a result of the rotational speed in the exhaust pressure chamber. Calculations show that this device will operate with an efficiency of about 80%; that is, the actual work output will be equal to 80% of the work output for a reversible adiabatic expansion.

Helical Rotary Screw Expander

At the time of this writing, Dr. Richard A. McKay and Mr. Roger S. Sprankle are evaluating this device based on developments already accomplished by Roger Sprankle and Hydrothermal Power Co. Ltd. (HPC). One objective is to determine whether or not additional development is warranted. The prototype machine was tested on the brine from East Mesa well 6-1 and gave efficiencies of 65 to 74%. These results were a preliminary demonstration attempt without any effort made to optimize the operating conditions. The accepted adiabatic efficiencies for these machines are in the range of 70 to 85% operating on gases. The following presentation by McKay and Sprankle (1974) describes the machine:

The helical rotary screw expander is a machine based upon development work conducted by Alf Lysholm in Sweden in the 1930's. By the 1950's Lysholm's machinery began to see extensive commercial application as a gas compressor. This application has had continued growth to the present date with installations involving 100 MW currently in operation. Early in 1971, HPC began development work in applying the Lysholm machine as a prime mover operating on geothermal hot water and brine. This development work has continued to the present.

The screw expander is a unique positive displacement machine, which bridges the gap between centrifugal or axial flow type aerodynamic machines and reciprocating positive displacement machines. It runs in a slower speed range without the high radial loads and balance problems characteristic of turbines.

As a geothermal prime mover, the helical screw expander is a total flow machine, which can expand directly the vapor that is continuously being produced from the hot saturated liquid as it decreases in pressure during its passage through the expander. The effect is that of an infinite series of stages of steam flashers, all within the prime mover. Thus, the mass flow of vapor increases continuously as the pressure drops throughout the expansion process, and the total fluid is carried all the way to the lowest expansion pressure. The process approximates an isentropic expansion from the saturated liquid line for the total flow. The expansion within the machine can be illustrated with drawings such as Figure 7.7. The geothermal fluid flows through the internal nozzle control valve and at high velocity enters the high-pressure pocket formed by the meshed rotors, the rotor case bore surfaces, and the case end face, designated by *A* in the two figures. As the rotors turn, the pocket elongates, splits into a *V*, and moves away from the inlet port to form the region designated by *B*. With continued rotation, the *V* lengthens, expanding successively to *C*, *D*, and *E* as the point of meshing of the screws appears to retreat axially from the expanding fluid. The expanded fluid at low pressure is then discharged into the exhaust port.

Performance tests of the 62.5-kVA prototype HPC geothermal power plant were performed August 21, 1974, on brine from Well 6-1 at the U.S. Bureau of Reclamation geothermal test facility on the East Mesa KGRA. The two tests gave results of 65 and 74% for the expander efficiency compared with an ideal machine. The test tesults should be considered preliminary since no attempt was made to optimize the operating conditions or the test set-up and because the load was operated outside its calibrated range.

Assuming adiabatic operation, the first law of thermodynamics gives an idential expression for the performance for this machine to that of a turbine, Equation 7.4, except that the efficiency of the turbine is replaced with the efficiency of the helical screw expander:

$$W_A = e_H(h_1 - h_{2R}) \tag{7.19}$$

Tests at the Lawrence Livermore Laboratory show that the engine efficiency is 50% (Weiss et al., 1975; Weiss, 1975). The fresh water

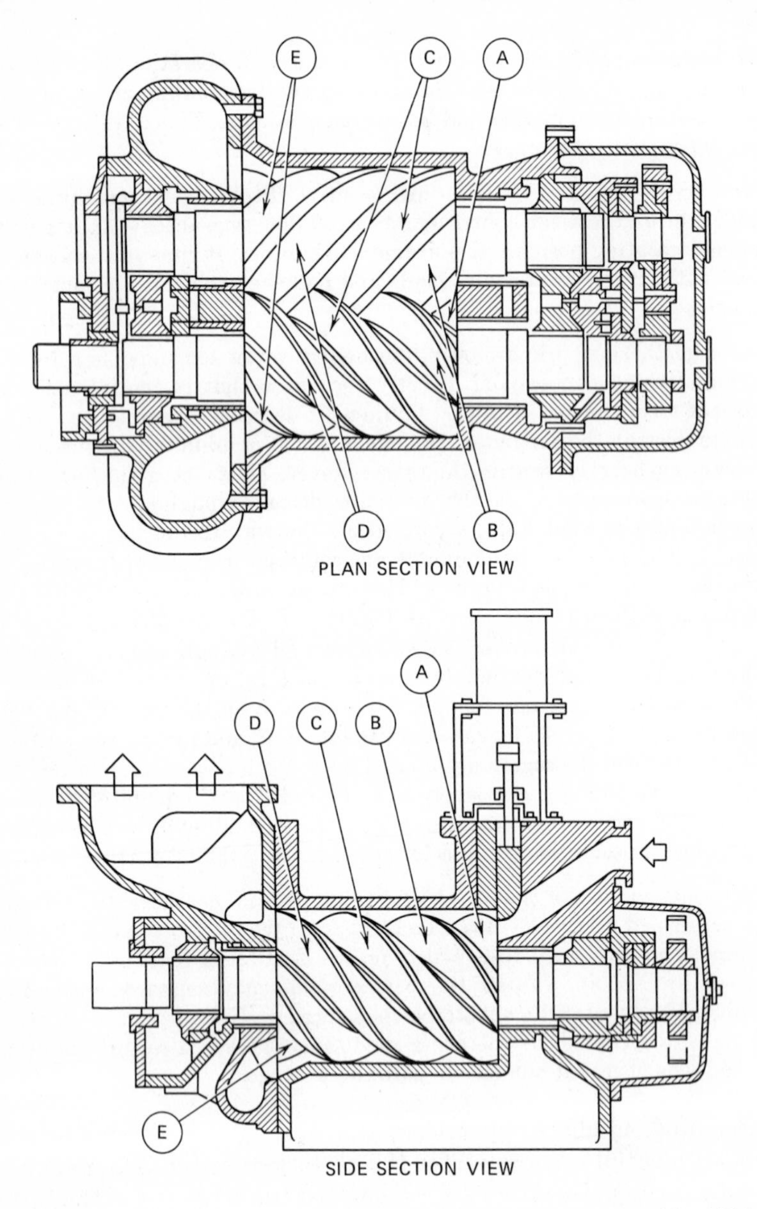

Figure 7.7 The helical rotary screw expander (McKay and Sprankle, 1974).

used in these tests dissolved some of the scale that was deposited in the machine during the earlier operation on geothermal brine and thus increased the leakage clearances within the machine. The effect of the resulting leakage on performance is not known (McKay, 1975). One disadvantage of this machine as a two-phase expansion device is the limitation of the pressure ratio over which it can operate. Because of its physical configuration, it is difficult to construct a machine for a pressure ratio much greater than 3:1. However, as a topping system as described in Chapter 9, this device would have excellent prospects.

Two-Phase Axial-Flow Impulse Turbine

This device, as being developed at Lawrence Livermore Laboratory (Austin, 1975), consists of a nozzle in which the fluid is expanded and a turbine wheel with blades to convert the kinetic energy of the exhaust fluid from the nozzle into rotating mechanical shaft work. This device is shown in Figure 7.8. The work output of the machine is given by the same equations as for other expansion machines, namely Equations 7.2 and 7.4, using the appropriate efficiency. For this machine, the efficiency will be the product of the efficiency of the nozzle for converting the fluid into kinetic energy, which is the same as that of

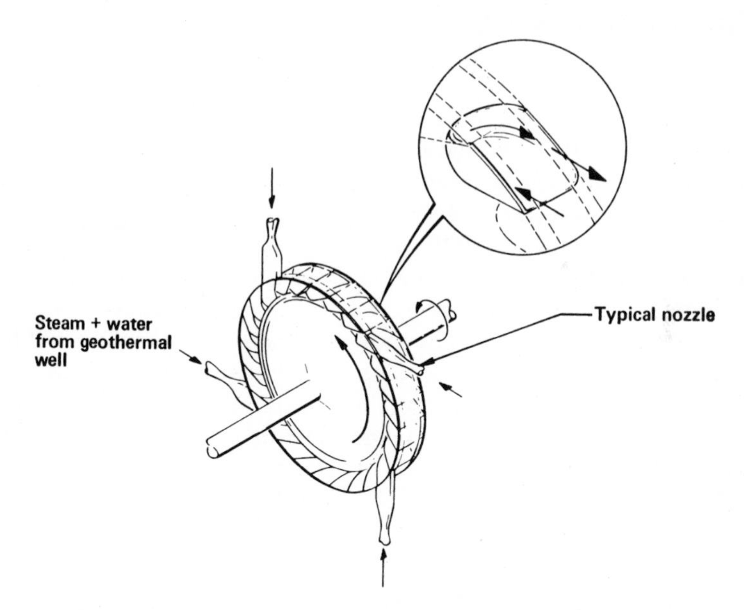

Figure 7.8 Two-phase axial-flow impulse turbine (Austin, 1975).

the nozzle machine described previously, times the efficiency of the impulse turbine blades for converting the kinetic energy into mechanical motion of the rotating wheel. As described for turbines, the efficiency of single-stage steam turbine blades will be 90%. Thus the overall efficiency must be less than the product of the two efficiencies, and this is 80%. More likely, the efficiency will be considerably less than this.

Two-Phase Rotary Separator Turbine Machine

This device (Hays and Elliott, 1975) is based on the high mass ratio of liquid to gas exiting from the two-phase expansion of a hot liquid in a nozzle. Since more than 80% of the mass, and typically as high as 90% for low pressure ratio expansion, is in the form of liquid water, most of the kinetic energy will be contained in the liquid stream. The key device in this system is a frictionless rotary wheel on which the expanded fluid impinges. Due to the centrifugal motion, the liquid is separated from the gas, which is then discarded. The high velocity liquid is then picked up by a reaction turbine or some other mechanical means that converts the kinetic energy of the liquid into mechanical work. Thus the basic thermodynamic relationship for this machine is the same as for a turbine or the device described previously. Specifically, the work output is given by

$$W_A = e_s(h_1 - h_{2R}) \tag{7.20}$$

The efficiency of this machine will be the product of the efficiency of the two-phase nozzle, the efficiency of the separator, the efficiency that represents the loss of kinetic energy in the gas stream, and the efficiency of recovery of energy from the high velocity liquid. The nozzle efficiency is 70 to 90% as in the previous devices, and the efficiency of the rest of the system is at best 80%. This system has the advantage of simplicty of design as does the nozzle reaction machine described previously. It could be built to handle a large volume of fluid with relatively small surface area and few moving parts. On the other hand, it consists of three units, the nozzle, the separator, and the power recovery unit, whereas a flash steam system consists of two units, the flash separators and the steam turbine.

SUMMARY

The thermal energy of a hot fluid is converted to mechanical work by expanding the pressurized fluid in a device that converts the kinetic energy of the expanding fluid into rotating shaft work. Expansion machines such as turbines have a high throughput relative to their size. Consequently, they operate adiabatically, that is without heat losses. Application of the first law of thermodynamics for an expansion process in which a fluid is expanding from a higher pressure and temperature to a lower pressure and temperature adiabatically is

$$W_A = h_1 - h_2 \tag{7.1}$$

If the expansion is a reversible one, that is, there are no friction losses, then the entropy change for the expansion process is zero. Thus the outlet enthalpy of the fluid is obtained by following a constant entropy path from the inlet condition:

$$W_R = (h_1 - h_{2R}) \qquad \text{for} \qquad \Delta S = 0 \tag{7.2}$$

This is the maximum work that can be obtained from any expansion process. Alternatively, the reversible work output can be calculated for the expansion of a gas from the pressure volume relationship of the gas according to Equation 7.11. The actual work is equal to the efficiency of the machine times the ideal reversible work output:

$$W_A = eW_R \tag{7.15}$$

A comparison of the devices as discussed in this chapter is summarized in Table 7.1. The overall efficiency is the product of the efficiencies of the expansion in the nozzle and the subsequent recovery of fluid kinetic energy as mechanical work. The efficiency of gas and steam

Table 7.1 Comparison of Two-Phase Flow Expansion Machines

	Efficiency			Pressure Ratio	
Device	Nozzle	Recovery	Overall	Near Opt. Eff.	Maximum
Nozzle-Reaction	0.9	--	0.9	10:1	100:1
Axial-Impulse	0.9	0.9	0.8	2:1	--
Helical Screw	--	--	0.5-0.7	3	5
Separated Phase	0.9	0.8	0.7	10:1	100:1

turbines varies from 50 to 85%. The efficiency of other machines presently under development for converting hot geothermal brine energy to mechanical work is not known with certainty, but the efficiencies of these machines are expected to be in the neighborhood of 50 to 80%.

NOMENCLATURE

a arbitrary constant
b_i arbitrary constants
e efficiency; e_i efficiency of the ith device, %/100
E internal energy, Btu
g conversion factor, 32.2 lb-m ft/sec^2 lb-f
h specific enthalpy, Btu/lb
H enthalpy, Btu
J conversion factor, 778 ft lb-f/Btu
m mass, lb
M molecular weight
n ratio of specific heats, c_p/c_v
P pressure lb-f/ft^2
Q heat, Btu
Q_m maximum heat, Btu
R gas constant, 1.987 Btu/lb mol °R
s specific entropy Btu/lb-°R
S entropy, Btu/°R
T absolute temperature, °R
U vapor velocity relative to nozzle, ft/s
u liquid velocity relative to nozzle, ft/s
U_{EX} exhaust velocity relative to earth, ft/s
V volume, ft^3
W specific work, Btu/lb
W_B specific work for a batch process, Btu/lb
X weight fraction
z compressibility (Weber and Meissner, 1959, Chapter 8)

Subscripts

A actual
EX exhaust conditions in condenser
H helical screw expander
M maximum, for a reversible process discharging at ambient temperature and pressure

n nozzle
0 ambient conditions
R reversible
s two-phase rotary separator expander
t turbine expander

REFERENCES

Alger, T. W., "The Performance of Two-Phase Nozzles for Total Flow Geothermal Impulse Turbines," Proceedings of the Second U.N. Symposium on the Development and use of Geothermal Resources, San Francisco, May 20–29, 1975.

Austin, A. L., "Prospects for Advances in Energy Conversion Technologies for Geothermal Energy Development," Proceedings of the Second U.N. Symposium on the Development and use of Geothermal Resources, San Francisco, May 20–29, 1975.

Hays, L. G., and D. G. Elliott, "Two-Phase Engine," U.S. Patent No. 3,879,949, 1975.

Marks, L. S., and T. Baumeister, *Standard Handbook for Mechanical Engineers, seventh ed.*, McGraw-Hill, New York, 1967.

McKay, R. A., and R. S. Sprankel, "Helical Rotary Screw Expander Power System," Proceedings on Research for the Development of Geothermal Energy Resources, Pasadena, Calif., September 23–25, 1974.

McKay, R. A., personal communication, 1975.

Perry, R. H., and C. H. Chilton, *Chemical Engineers Handbook, fifth ed.*, McGraw-Hill, New York, 1973.

Schultz, J. M., "The Polytropic Analysis of Centrifugal Compressors," *Trans. Amer. Soc. Mech. Eng.*, 69–82 (January 1962).

Spencer, R. C., et al., "A Method for Predicting the Performance of Steam Turbine-Generators," *J. Eng. Power*, 254 (October 1963).

Swearingen, J. S., "Turboexpanders and Processes That Use Them," *Chem. Eng. Progr.*, 68, 95–102 (1972).

Weber, H. C., and H. P. Meissner, *Thermodynamics for Chemical Engineers*," Wiley, New York, 1959.

Weiss, H., et al., "Performance Test of a Lysholm Engine," Lawrence Livermore Laboratory, Univ. California, Livermore, prepared for U.S. Energy Research & Development Administration under contract No. W-7405-Eng-48, 1975.

Weiss, H., "Two-Phase-Flow Test Facility for Geothermal-Energy Conversion Machinery," Lawrence Livermore Laboratory, Univ. California, Livermore, prepared for U.S. Energy Research & Development Administration under contract No. W-7405-Eng-48, 1975.

CHAPTER 8

Electrical Power Production Using Expansion Machines

Electrical power production plants for utilizing geothermal energy are brine processing plants that extract the thermal energy of the hot brines and convert it into electrical energy. The best process plant design for any of the expansion machines described in Chapter 7 will minimize irreversible losses such as mechanical and fluid friction and temperature differences in heat transfer processes. If the plant is perfectly designed and operated so that all such irreversible losses are negligible, then, as was shown in Chapter 6, no matter what the system is nor what the expansion machine or components within the system, the work output will be

$$W_R = (h_1 - h_2) - T_{EX}(s_1 - s_2) \tag{8.1}$$

Since all real process plants will operate with some irreversible losses, the actual work output will be less than given by this equation. As shown in Chapter 2, most geothermal brines are relatively dilute and have thermodynamic properties very close to those of pure water. Even brines with very high total dissolved solids, such as the Salton Sea brines, will have thermodynamic properties not very different from those of water. Consequently, an understanding of the comparative performance of various expansion machines and process plants can be developed using the thermodynamic properties of water to represent those of the geothermal brines to be exploited. Thus the graph of the available work function, Equation 8.1, for pure water as a function of temperature given in Figure 6.5, represents the maximum work that can be obtained from a hot brine by a well-designed process plant.

PERFORMANCE OF REVERSIBLE PROCESS PLANTS

As described in Chapter 6 and shown schematically in Figure 6.1, there are two methods for converting thermal energy into mechanical or electrical work. One method is to expand the pressurized fluid through an expansion machine. The other method is to transfer the heat from the geothermal fluid to a secondary pressurized working fluid, which is then expanded to produce mechanical or electrical work.

For both methods of conversion, Equation 8.1 applies. Furthermore, from the definition of entropy, Equation 6.7, the heat exhausted from the process is given by the second half of Equation 8.1 and is

$$Q_{EX} = T_{EX}(s_1 - s_2) \tag{8.2}$$

This is the heat that is removed by a condenser. The condenser determines the condensation temperature and therefore exhaust pressure from the expansion machine. To obtain the maximum work from the expansion machine, the condenser should operate at the lowest available temperature of the environment, so that $T_{EX} = T_0$. For a system in which the thermal fluid is pressurized and then expanded, Equation 8.1 gives the net work output.

If the heat is transferred to a secondary working fluid, a further basic relationship can be derived relating the temperatures at which the heat is exchanged to and from the working fluid system, known as the Carnot cycle, outlined in Figure 8.1. In this system, heat is transferred from a thermal reservoir reversibly at the temperature T_1 to the secondary working fluid system, and heat is rejected from the system at a lower temperature equal to the temperature of the environment. The application of the first law of thermodynamics to the secondary working fluid system shown by the dotted boundary line in Figure 8.1 is

$$W_R^* = Q_H - Q_C \tag{8.3}$$

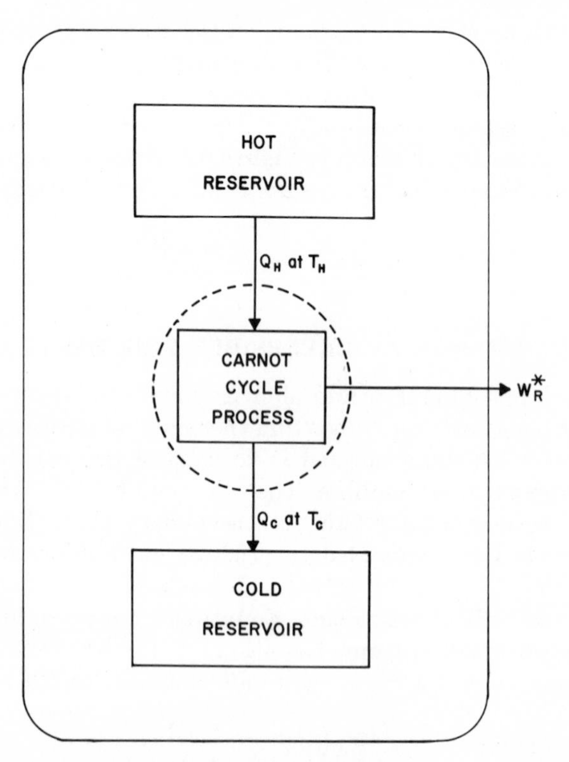

Figure 8.1 The Carnot cycle process. A schematic representation of a reversible process for transferring heat from a hot to a cold source reversibly and producing work.

For the entire system enclosed by the solid boundary in Figure 8.1, there is no heat flow in or out. Consequently, according to the definition of entropy, Equation 6.7,

$$\Delta S = \frac{Q_R}{T} = 0 \tag{8.4}$$

Since the entropy change of the entire system is zero and because the secondary working fluid system will be operating at steady state and there will be no accumulation or depletion of entropy within that system, any entropy change in the hot reservoir must be equal to the negative of any entropy change in the cold reservoir. Consequently,

$$\Delta S_{RH} = -\Delta S_{RC} \tag{8.5}$$

Since the heat flows are all reversible, according to the definition of entropy, Equation 6.7,

$$Q_H = -T_H \Delta S_{RH} \tag{8.6a}$$

$$Q_C = T_C \Delta S_{RC} \tag{8.6b}$$

Combining Equations 8.5 and 8.6 gives

$$\frac{Q_H}{Q_C} = \frac{T_H}{T_C} \tag{8.7}$$

Inserting this value of the ratio of the heat flows in Equation 8.3 results in the familiar relationship for the work obtained from a reversible secondary fluid working system called the Carnot cycle:

$$\frac{W_R^*}{Q_H} = \frac{(T_H - T_C)}{T_H} \tag{8.8}$$

This equation gives the efficiency with which thermal energy available from a heat source can be converted into work. A plot of this equation as a function of the temperature of the heat source, Figure 8.2, shows that the efficiency falls rapidly with decreasing temperature of the heat source. Since geothermal energy sources are generally available in a temperature range of 300 to 500°F, the rejection temperature available for the condenser has a significant effect on the efficiency of the system. Figure 8.3 presents the relationship between work output and the rejection temperature for various hot reservoir temperatures. These figures represent the work output that is obtained from a reversible process that is absorbing heat into the system at the temperature T_H and rejecting it to the environment at some temperature T_C.

The maximum energy will be absorbed from a geothermal brine by cooling the brine from its wellhead temperature to the ambient temperature of the surroundings. To do this reversibly with a Carnot system as described already, the heat is transferred in small increments to a series of systems each of which are receiving the heat at the temperature of the brine. This is shown in Figure 8.4. For such a system, the net work output will be the sum of the work outputs of all the individual systems. For an infinite number of infinitesimally small Carnot engines, this is

$$W = \int_{T_1}^{T_2} \left[\frac{T_H - T_0}{T_H}\right] dQ_H \tag{8.9}$$

where Q_H is the heat flow for 1 lb of brine. For this limiting case, the work output is also given by Equation 8.1, so that Equations 8.1 and 8.9 will yield identical results.

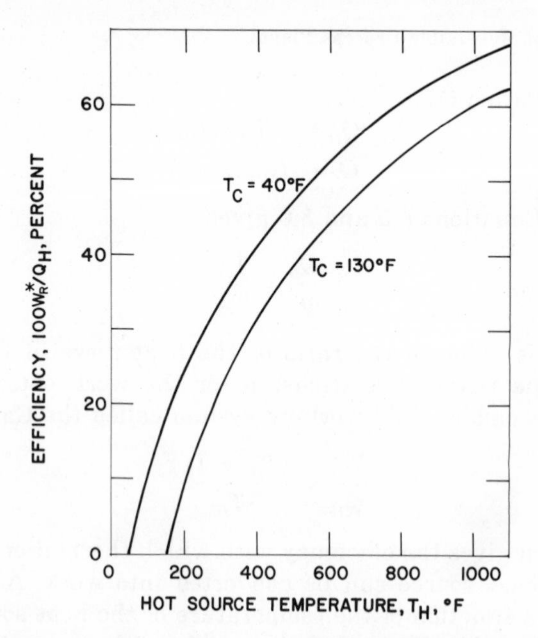

Figure 8.2 The efficiency, that is, the fraction of the thermal energy that is converted to work by a reversible process, the Carnot cycle process, as a function of the temperature at which heat is being delivered to the reversible process. Note the significance of the temperature T_C at which heat is being rejected by the process for lower hot source temperatures.

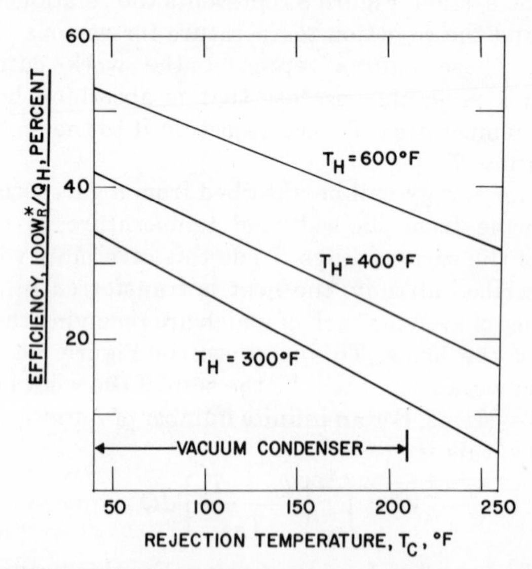

Figure 8.3 The efficiency for the reversible conversion of heat to work by the Carnot cycle process as a function of the rejection temperature of the process for various hot source temperatures.

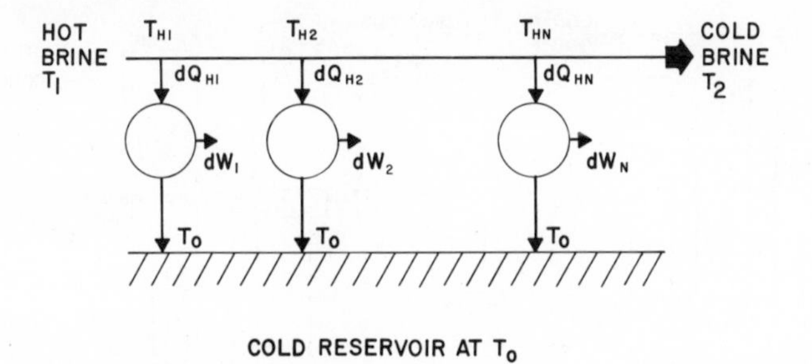

Figure 8.4 The reversible recovery of heat from a geothermal brine to produce work by a series of infinitesimally small Carnot cycle processes that are discharging the waste heat to a cold reservoir that is at ambient temperature.

PERFORMANCE OF REAL PROCESS PLANTS

The important question with respect to the use of Equation 8.1 and Figures 6.5 and 6.6 is how close do actual process plants come to this ideal performance? The technological development of plant system designs and process operations has reached a high degree of perfection. The plants approach very closely reversible performance, providing appropriate allowance is made for the limiting efficiency of the basic pieces of equipment used in the process.

The Etiwanda electrical power generating plant of San Diego Gas and Electric Co., shown schematically in Figure 8.5 and described in Table 8.1, is an example of such a case. The overall plant thermal efficiency is 37%. For this plant, the Carnot cycle efficiency , which is the work obtained from a reversible process operating between the boiler temperature of 1000°F and the condenser temperature of 100°F, is 61.7%. As shown in Chapter 7, the best efficiency that has been obtained from turbine machines is 85% and represents the limit of existing technology. The efficiency of electrical generators is about 97%. The boiler efficiency for this plant, which is the measure of the amount of thermal energy transferred from the combustion stream to the hot water in the boiler, is reported to be 88%. The product of these three efficiencies and the Carnot cycle efficiency gives an overall plant efficiency of 45%. This means that all of the other losses in the system must amount to no more than 8%. Consequently, this plant design and operation has losses amounting to less than 8% of the limiting work output using the best available expansion turbine, electrical generator, and boiler. However, if better materials of construction were available or different working fluids were used so that the boiler

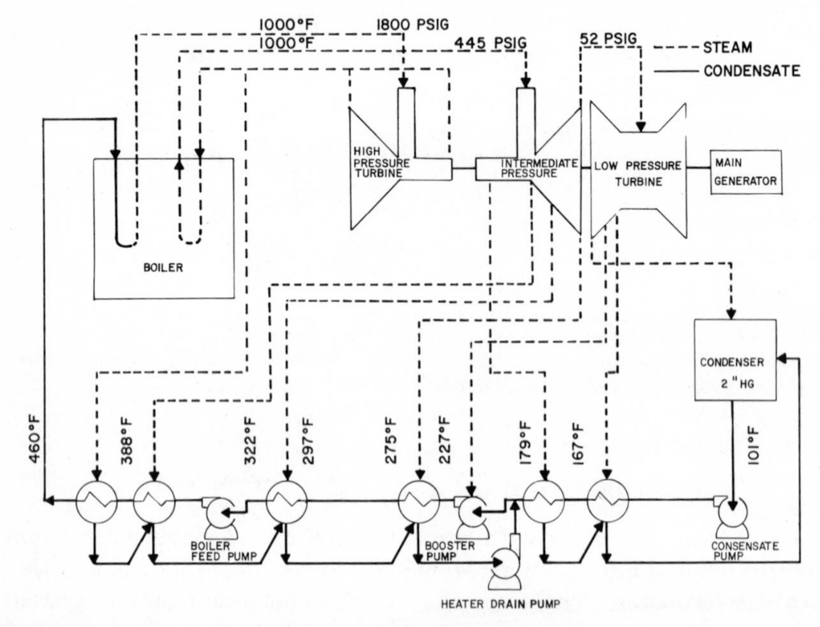

Figure 8.5 Schematic representation of the Etiwanda power plant. Note the countercurrent heating of the condensate up to a temperature of 460°F. This is accomplished by drawing off steam at various temperatures in the expansion process for heating the boiler water. Also this condensate flows countercurrent to the heat exchangers as shown in the lower part of the diagram.

Table 8.1 Operating Characteristics of One Turbine Unit of the Etiwanda Power Plant Shown in Figure 8.5

Boiler	1000°F
Condenser	100°F
Power Output	260 MW
Oil Consumption	17,000 gal/hr
Boiler Efficiency	88%
Carnot Cycle Efficiency	61.6%
Thermal Efficiency	37%

and turbine inlet could be operated at higher temperatures, then a greater power output would be obtained because the Carnot cycle efficiency would be higher. Combustion gases produce temperatures approaching 3000°F and so considerable work can be obtained from a so-called topping cycle. A topping cycle is a power plant cycle, such as one using mercury as a working fluid, to produce work from the heat flow from the combustion temperature to the 1000°F receiving temperature of conventional plants. Alternatively, gas turbines have been suggested for this purpose.

This discussion shows that for systems operating below 1000°F and using the best turbine and generator available with today's technology, a system can be designed and operated so that all other irreversible losses are small compared to the total work output. Of course, to install all of the equipment necessary to keep the irreversible losses to a minimum, a large power plant may be necessary. Such power plants may be larger than is economically feasible considering the problem of collecting the thermal energy from a number of scattered wells. Thus the most economical geothermal power plant may not be designed to the same state of perfection as large fossil fuel power plants.

GEOTHERMAL WELL OUTPUTS

The largest practical diameter of wells with existing technology is between 6 and 10 in. Consequently, the flow rates of geothermal brine–steam mixtures from geothermal wells is limited by the throughput of a conduit of this diameter. As the brine flows up the well from the reservoir, the hydrostatic pressure drops due to fluid friction losses. Consequently, the higher the flow rate the lower must be the wellhead pressure to maintain this flow rate. The relationship between flow rate and wellhead pressure is typified by the characteristics of the Otake geothermal well output versus pressure data shown in Figure 8.6. Note that this data shows there is a limit to the flow rate that can be achieved by dropping the pressure. This is due to sonic choking of the well; that is, the velocity of the steam approaches sonic velocity in the well thus limiting the flow.

The wellhead product from a geothermal reservoir may be pure steam in the case of steam-dominated fields or it may be pure water in the case of low temperature reservoirs with sufficient pressure to move the fluid to the wellhead. The most common geothermal reservoir will

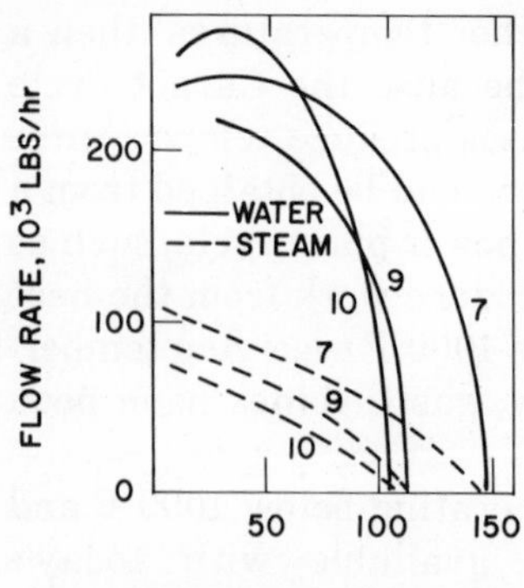

Figure 8.6 Typical pressure flow rate relationships for geothermal wells. The total flow consists of steam flow plus water flow. Note that the proportion of steam increases with decreasing pressure and increasing flow rate. This data is for geothermal wells at Otake in Japan, the numbers on the curves being the well numbers. From Usui and Aikawa (1970).

probably deliver a mixture of steam and water to the wellhead depending on the temperature and pressure conditions of the reservoir and the flow rate that is set by the pressure at the wellhead. Wells in steam-dominated fields typically produce 50,000 to 150,000 lb of steam/hr. Wells in hot brine reservoirs will typically deliver combined brine and steam flow rates in the neighborhood of 50,000 to 500,000 lb/hr. In either case the output of a single geothermal well contains sufficient thermal energy to produce somewhere between 1 and 10 MW of electrical power. Since this power level is much too low for the smallest reasonably efficient and economic steam turbine, the fluid output from a number of wells must be gathered to power a single turbine generating unit. Typically, somewhere between 5 and 20 wells will be gathered together by a pipe transmission system to feed one turbine–generator unit.

FLASH STEAM GEOTHERMAL ELECTRICAL GENERATING PLANT

The simplest scheme for generating electricity from a steam dominated field is to remove entrained moisture and particulate matter from the wellhead steam and expand the steam through a turbine. The corresponding method for hot brine reservoirs is to separate clean steam from the steam and brine mixture. This clean steam is then passed through a turbine to generate power. In the simplest case, a single separator and turbine unit may be operated somewhat below wellhead pressure. This lower pressure is used to produce a greater quantity of flash steam at a pressure and temperature slightly below wellhead conditions, thus giving a greater power output from a given well capacity.

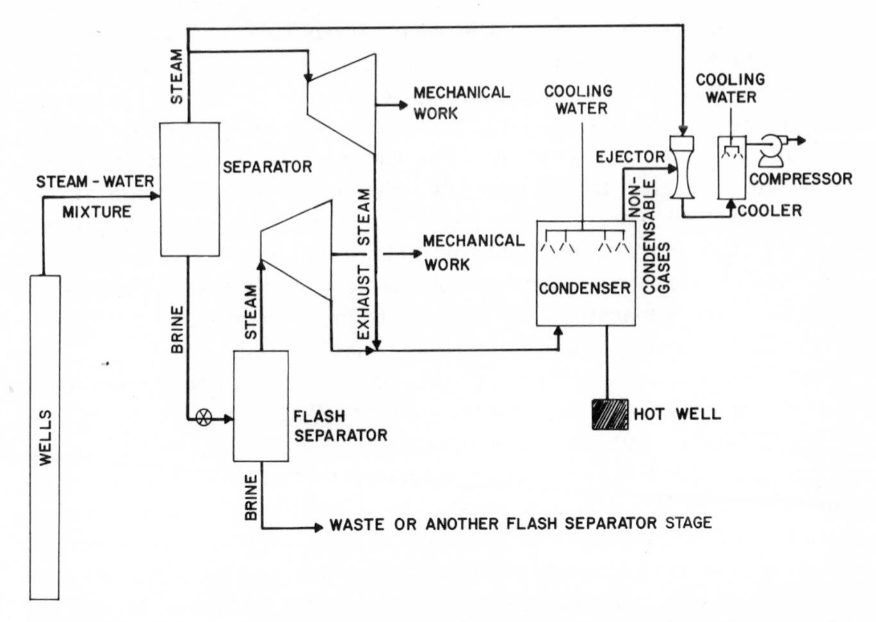

Figure 8.7 Process schematic for the generation of power from geothermal steam water mixtures by the two-stage flash separator system. Note the noncondensable gas ejection system for the condenser. For geothermal brines the power requirement for removal of noncondensables is a significant factor. Note that although two turbines are shown, in actual practice only one turbine might be used with the low pressure steam injected at an appropriate location in the turbine.

The more recent geothermal power plant installations are designed to operate with a two-stage flash system as shown in Figure 8.7. Since geothermal brines contain a substantial quantity of noncondensable gases, it is necessary to operate a gas compression system for removing these from the condenser as shown in Figure 8.7. Typically, the power requirements for the noncondensable gas removal is about 2 MW for a 50-MW power plant. As shown in Figure 8.3, considerably greater power output can be obtained by exhausting the turbine to a condenser that is producing a vacuum. The exhaust suction that can be maintained by the condenser is determined by the available cooling water temperature. For northern climates, this temperature is much lower than it would be, for example, from cooling towers operated in a desert or tropical region. The range of possible condenser temperatures varies from 40 to 130°F.

As shown in Figure 8.7, the steam–water mixture from a geothermal well is first separated so that the steam can be piped directly to a turbine. The hot brine from the separator is then passed through a flash separator where a pressure drop allows some of the water in the brine to flash. This vaporization causes a drop of the brine temperature. The steam from the flash separator is then passed through a low pressure turbine, and the work is extracted. In actual practice, a single turbine may take the high pressure steam at the inlet and the lower pressure steam from the flash separator at an intermediate stage to produce the mechanical work. The brine from this flash separator, as shown in Figure 8.8, could then be passed to another flash separator and another turbine to produce additional work. The number of such flash stages that would be economic would depend on the temperature of the reservoir. Higher temperature reservoirs would allow the economic use of a larger number of stages.

The power output from a brine–steam system such as is shown in Figure 8.7 or 8.8 can be calculated based on the thermodynamics of the system. In the separator stage, 1 lb of wellhead fluid consisting of brine and steam is separated as shown in Figure 8.9 to produce x_1 lb of brine at the pressure and temperature entering the separator and $1 - x_1$ lb of steam at the same pressure and temperature. This brine

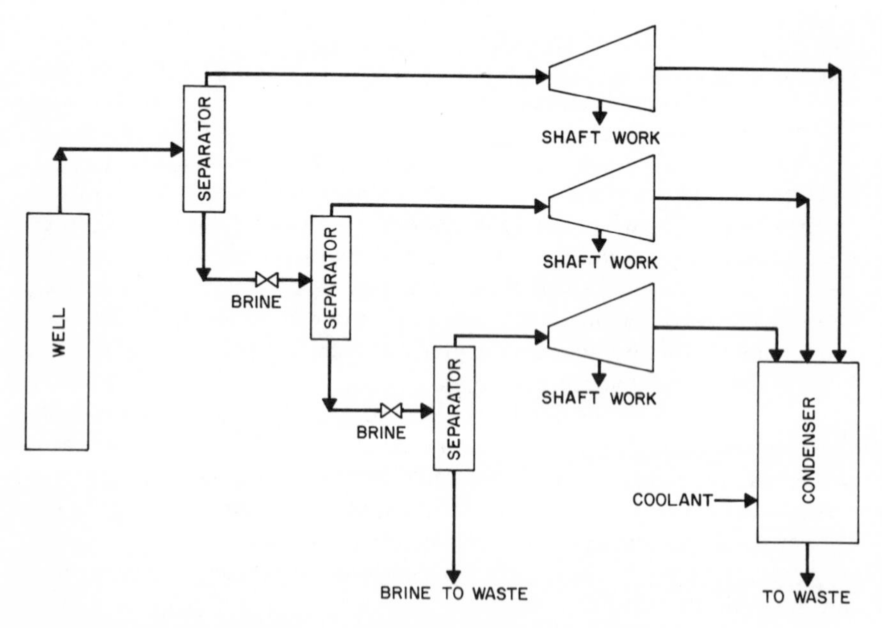

Figure 8.8 Power production from geothermal brine by expanding steam that has been separated from brine flashed in a multiple set of flash separators.

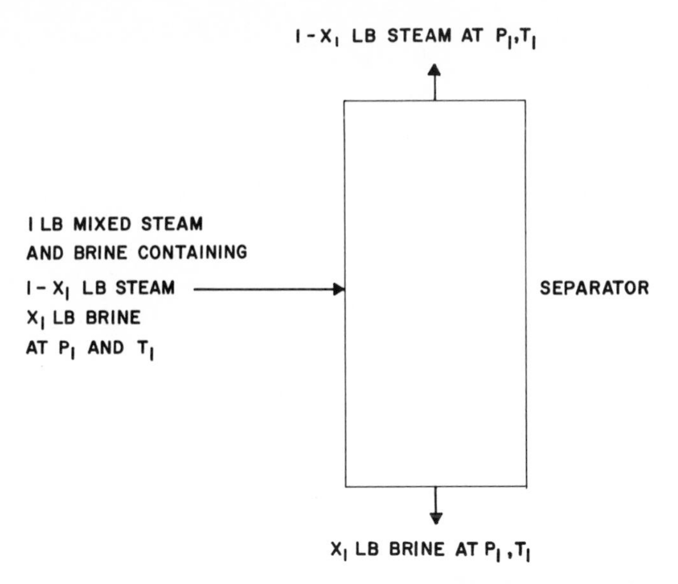

Figure 8.9 Process schematic for the first-stage separator that separates the steam and brine being delivered from the well transmission system.

may then be fed through a pressure drop regulating valve to a second separator, Figure 8.10, in which flashing occurs because of the decreased pressure. Consequently, it is called a flash separator. The product is x_2 lb of brine and $x_1 - x_2$ lb of steam, both at the lower pressure P_2 and temperature T_2. The temperature of the brine and steam and the amount of steam produced is a function of the pressure at which the flash separator operates and of the wellhead temperature and pressure. This relationship is determined by writing an energy balance around the flash separator that results in the following relationship between the enthalpies and mass flow rates:

$$x_1 h_{L1} = x_2 h_{L2} + (x_1 - x_2) h_{V2} \tag{8.10}$$

This equation can then be solved to give the ratio of the output to input brine flow rates:

$$\frac{x_2}{x_1} = \frac{(h_{V2} - h_{L1})}{(h_{V2} - h_{L2})} \tag{8.11}$$

and the ratio of steam produced to the input brine:

$$\frac{(x_1 - x_2)}{x_1} = \frac{(h_{L1} - h_{L2})}{(h_{V2} - h_{L2})} \tag{8.12}$$

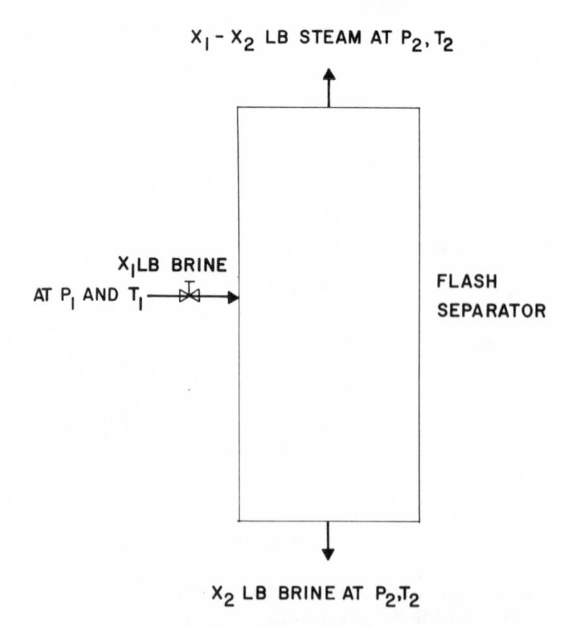

Figure 8.10 Process schematic for a flash separator. In a flash separator, the pressure of the brine drops so that some steam vaporizes and the temperature of the brine drops accordingly. The separator then separates flash steam from the brine. The net result is the production of steam as a result of dropping the pressure of the brine.

The work produced by a turbine using steam from the first and second stage separators can be calculated for the conditions shown in Figures 8.9 and 8.10 using Equation 7.4:

$$W_1 = e_{t1}(1 - x_1)(h_{V1} - h_{EX}) \qquad \text{for} \qquad \Delta s = 0 \qquad (8.13a)$$

$$W_2 = e_{t2}(x_1 - x_2)(h_{V2} - h_{EX}) \qquad \text{for} \qquad \Delta s = 0 \qquad (8.13b)$$

and for the ith stage by

$$W_i = e_{ti}(x_{i-1} - x_i)(h_{vi} - h_{EX}) \qquad (8.13c)$$

The enthalpy change for this equation is calculated assuming a reversible adiabatic expansion. The efficiency of turbines used for geothermal steam will have a rather low efficiency since they are low pressure turbines. As shown by Figure 7.4, this efficiency will be about 70%. Adding up the work output for each of the turbines for each of the stages will give the total work output for the plant. From this work output must be deducted the work for removal of the noncondensable gases as well as any pump work necessary for reinjecting the spent brine and circulating cooling water. This power requirement will be in the neighborhood of 1% of the total power output of the plant. The power output from a multiple-flash plant neglecting heat losses and pump work and assuming a turbine efficiency of 70%, is shown in Figure 8.11 for various numbers of flash stages. The power output depends on the discharge pressure of the turbine because of the Carnot cycle efficiency–temperature relationship discussed previously. Consequently, the power output of the multiple-flash system is shown for several discharge pressures for a reasonable number of flash stages in Figure 8.12. For a turbine efficiency of 100% and an infinite number of flash stages, the steam flash power plant will produce an amount of work equal to the limiting work of a reversible process given by Equation 8.1 and shown in Figures 6.6 and 8.11.

The optimum pressure of the second stage for a two-flash power plant system is only slightly above atmospheric pressure. The optimum pressure of the first-stage separator is about midway between the wellhead pressure and the second-stage pressure. The ratio of the temperature drop of the second-stage flash to the first-stage flash for maximum work is 1 to 1.3. In the case of more than two flash stages,

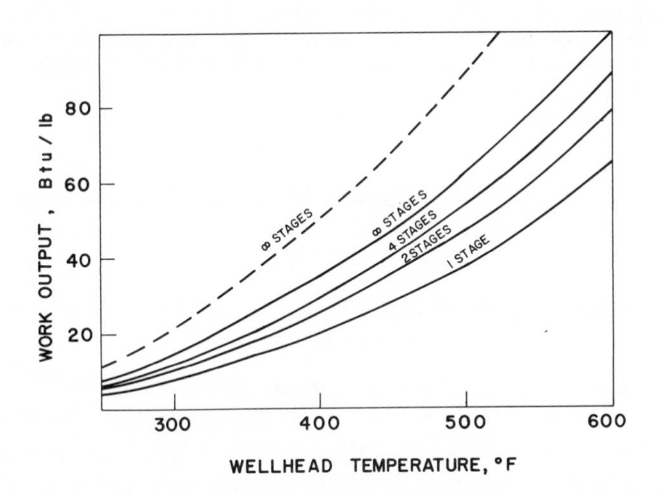

Figure 8.11 The work produced by a flash steam geothermal power plant with a turbine efficiency of 70% and an exhaust pressure on the turbines corresponding to a condenser temperature of 130°F. The work is shown for various numbers of flash stages as indicated by the numbers on the curves. The outlet to inlet temperature ratio for each flash separator is 0.5. The dashed line in the figure is the work obtained for a 100% efficient turbine. The thermodynamic properties of the brine are assumed to be the same as those of water.

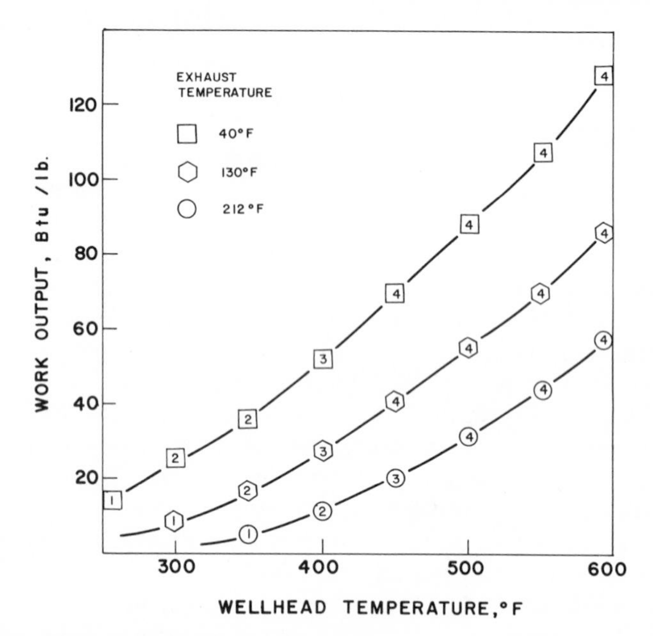

Figure 8.12 The work produced per pound of brine by a multistage flash geothermal power plant for various exhaust pressures and condenser temperatures. It is assumed the thermodynamic properties of the brine are those of water. A turbine efficiency of 70% was used. The numbers inside the data points indicate the number of flash stages that were assumed. For the higher temperatures a larger number of stages is used because the higher temperature allows economic use of a larger number of flash stages.

the optimum temperature drop ratio between successive stages is 1 to 1.3.

Capital Cost of Geothermal Flash-Steam Powered Electricity Generating Plants

The cost of electric generating systems using steam from a steam dominated field or steam flashed from the hot brine of a hot water reservoir is about $300 per installed kilowatt. This cost includes the cost of development and production of the reservoir and the cost of the plant for converting steam into electricity as shown in Table 8.2. These cost figures generally apply over the range of sizes of geothermal power plants, which, as shown by the data in Figure 8.13, suggests an installed cost for a power plant in the order of $160/kW. The cost for both hot water- and steam-dominated fields is about the same because the cost of producing steam from hot water is small compared

Table 8.2 Range of Capital Costs and Operating Costs in 1975 Dollars Per Kilowatt for Geothermal Steam or Hot Water-Flashed Direct Steam-Driven Electricity-Generating Plants Based on a 110 MW-Sized Plant

Capital Costs Per Kilowatt		
Power Plant		
Turbogenerator	$40 to 70	
Electric Equipment	30 to 40	
Other	55 to 90	
Total Power Plant		$125 to 200
Collection System Piping		12 to 30
Wells at 80% success excluding royalties		100 to 150
Total Capital		$237 to 380
Annual Costs Per Kilowatt		
Well Replacement And Effluent Disposal		$ 20 to 40
Fixed Charges at 10% of Capital		24 to 38
Operation & Maintenance, Power Plant		6 to 10
Total		$ 50 to 88
Annual Cost Per KW-HR		5.8 to 10 mills

Source. Prepared from data of Burr, 1975, Bloomster, 1975, and Futures Group, 1975.

to the total plant cost. As noted before, the power capacity of steam wells and hot water wells is about the same. As shown in Table 8.2, the total capital cost of a geothermal power plant facility is split about equally between the power plant itself and the cost of the wells and the collection and transmission facilities. Thus costs can be significantly reduced by maximizing the power output obtained from each well. This implies maximizing the thermal output from each well, that is,

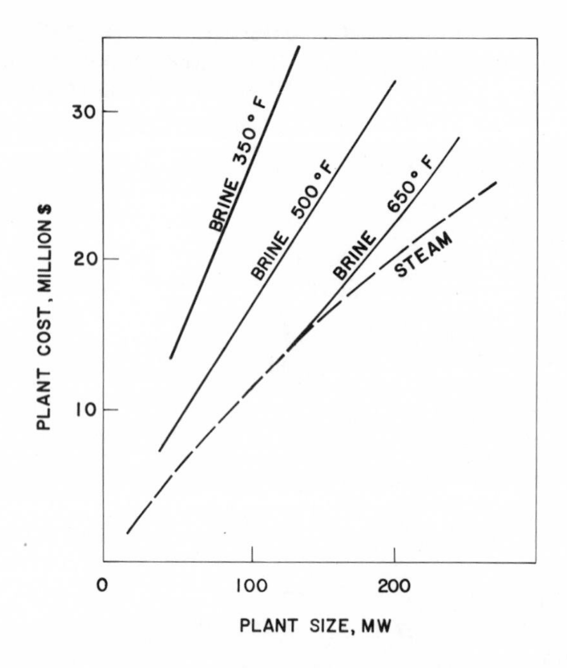

Figure 8.13 Cost in 1975 dollars of steam turbine geothermal power plants for steam dominated and hot brine reservoir systems as a function of the installed capacity in megawatts. Based on data from Bloomster (1975), Bloomster et al. (1975), Towse (1975), Barr (1975), Greider (1974) and the Futures Group (1975). The brine plant costs are based on cost data for Cerro Prieto, Wairakei, and Otake, which have reservoir temperatures between 475 and 575°F and wellhead temperatures of 450 to 550°F. The cost for plants with wellhead brine temperatures as shown was obtained by scaling the cost data for the process portion of the plant by the 0.6 power of the brine rate and assuming the electric portion was independent of the brine rate.

the product of the mass flow rate and the specific thermal energy content, as well as maximizing the conversion of this available energy into electricity in the generating facility.

Since the operating costs of geothermal power plants are low relative to the capital costs, the cost of the electricity produced from geothermal power plants is determined primarily by the capital cost of the geothermal installation. Thus a minimum annual cost per kilowatt corresponds to a minimum capital investment per kilowatt. For geothermal power plants at the present stage of technology, this minimum occurs for power plants of 100 MW. This is a result of the number of wells and the collection facilities for the power plant. The operating costs are 10 to 20% of the total electrical production costs for a 100-MW power plant. The 1975 cost of electricity produced is 5 to 10 mil/kWh and generally close to 7 mil/kWh for most systems.

The optimum plant size will depend on the relationship between the capital costs of the plant and the capital investment in the reservoir steam production system. This must be determined for each field. Wells may have to be replaced at a rate as high as 10%/yr. Thus the cost of the production facilities will increase with power plant size not only because of the greater distance that the brine must be transmitted to the generating facility but also because of the replacement of lines and production of new wells in more remote locations. As shown in Figure 8.14, the minimum capital investment will be a result of a

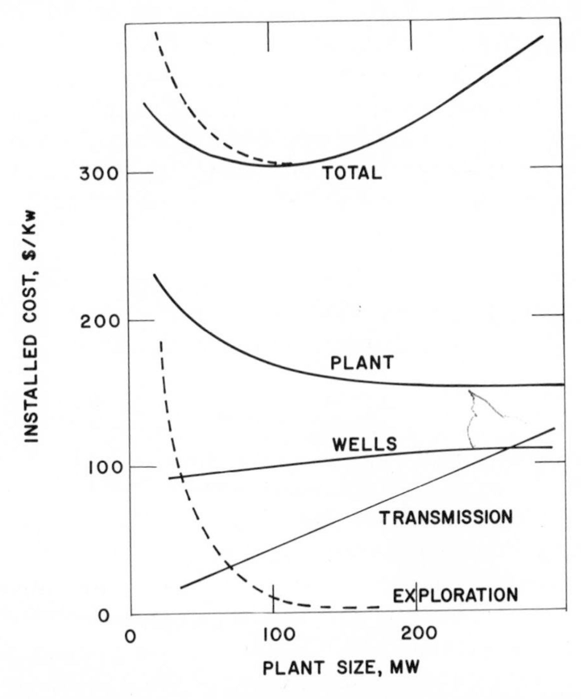

Figure 8.14 The relationship of the major cost elements of a geothermal power electricity generating plant system to plant size. Well costs tend to increase with size because of the use of deeper and less productive holes. Transmission costs include the collection system and the injection system, which increases with size because of the greater transport distance. The total cost does not include the cost of development, leases, and royalties. The inclusion of these costs would shift the optimum plant size to a higher level. Of course, the actual plant costs and optimum size will depend on the location and nature of the geothermal field as well as other local factors.

proper balance between generating facility costs and the size of the reservoir production system and transmission costs.

Because of the importance of the conversion efficiency of the available thermal energy to electricity, and because geothermal power plants are in their initial stages of development, improvement in efficiency and reduction of cost can be expected in the near future. Although little can be done with steam-dominated fields, an improvement in conversion of hot brines to electricity can be accomplished by using multiple-stage flash systems as opposed to the single-stage flash systems used in earlier plants. As shown in Figure 8.11, geothermal power plants with one flash stage operate at about 40% of their potential. The more sophisticated modern fossil fuel-powered electricity-generating plants operate at better than 50% of their potential 60% efficiency. It is reasonable to expect that further sophistication in geothermal power plant design will result in increased efficiency and reduced cost of the produced electricity. Since well costs and transmission costs are a dominant factor, it would be expected that improvement in well design, reservoir production, and transmission technology would also result in a significant reduction in cost of the electricity produced from a geothermal field.

COMPARISON OF SYSTEMS USING VAPOR- AND LIQUID-DOMINATED RESERVOIRS

The maximum power output from a plant using vapor dominated reservoirs is obtained by efficiently collecting and transmitting the steam to the power production facilities and efficiently expanding it to the pressure of a condenser operating at the lowest temperature of the surrounding environment. On the other hand, maximizing the power output from a liquid or mixed liquid and steam reservoir becomes complicated. In the case of mixed steam and brine delivered at the wellhead, the maximum power will be obtained by separating the steam and brine without pressure drop and then operating enough flash separator stages to obtain almost all of the available work output. As an example, the work output and efficiency as a function of the steam–brine ratio is shown for various numbers of flash stages in Figure 8.15.

An important thermodynamic loss to be minimized by good system design is the irreversible loss resulting from a large temperature drop between flash stages. If the brine entering a separator is at a much higher temperature than the temperature of the produced steam

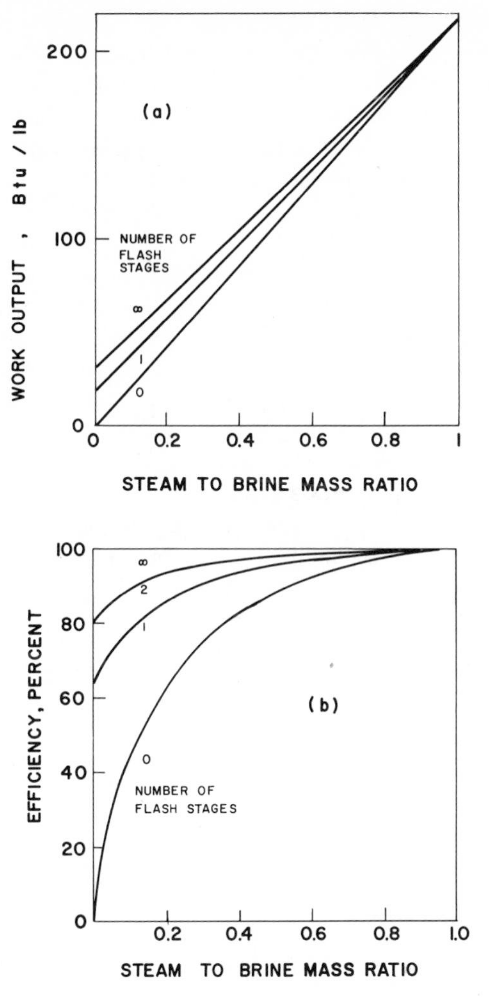

Figure 8.15 Power production as a function of the ratio of steam to brine delivered to the power plant by the reservoir system. The thermodynamic properties of water have been used. A condenser temperature of 130°F, a wellhead temperature of 400°F, and a turbine efficiency of 70% has been assumed. The efficiency loss with increasing steam-to-brine ratio is due to the irreversible heat losses as a result of temperature differences in the flash separator operation. (*a*) Work output. (*b*) Efficiency expressed as the ratio of the power output to the power output for an infinite number of flash stages.

leaving the separator, there is a large irreversible thermodynamic loss due to the large temperature drop. As shown by Figure 8.11, power output is maximized in the flash steam system by increasing the number of flash stages so that the irreversible heat losses, that is, temperature drop between each stage, is a minimum. As the number of flash stages approaches infinity, the temperature drops approach zero, and the power output approaches the limiting power output capability of the plant.

There are two primary thermodynamic losses in the power plant system just described. One is a loss resulting from low turbine efficiency, and the other is a loss due to the irreversible temperature effects in the flash separators. The significance of these factors is shown in Figure 8.16 in which a comparison is made between a two-stage flash system and an infinite number of flash stages. Also, the output for a plant operating with 70% turbine efficiency is compared with one operating at 85% turbine efficiency. The gain in efficiency resulting from these two effects gives an increase in power output of about 60%, which would result in nearly a 40% reduction in the cost of the electricity produced. The binary or secondary working fluid power plant system and the two-phase power plant systems described later

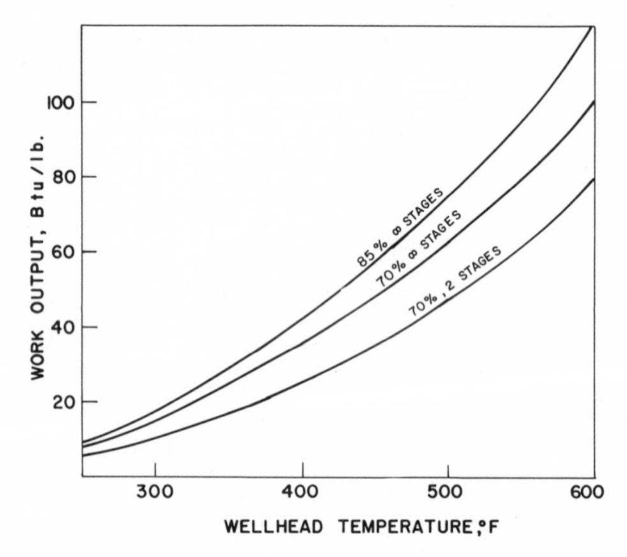

Figure 8.16 Power output as a function of brine temperature of a multistage flash separator system for various turbine efficiencies and number of stages as indicated on the curves. The thermodynamic properties of water are assumed. The condenser temperature is 130°F.

overcome these difficulties, so that potentially they could accomplish this improvement. However, they have other inefficiencies so that this cannot be easily achieved.

SECONDARY WORKING FLUID SYSTEMS APPLIED TO STEAM-PRODUCING SYSTEMS

Steam is produced from geothermal reservoirs either directly in the case of a vapor-dominated field or by flashing brine. Once this steam is produced, the most efficient process for converting it to mechanical work is probably the expansion of the steam directly in a turbine. The use of a secondary fluid in a power plant using the steam as the power source has been proposed. The only thermodynamic advantage that can be gained by such a system is the increase in efficiency of the turbine as a result of the higher working pressure of the secondary fluid system. Corresponding to the higher working pressure of the system would also be a larger plant capacity to take advantage of the increased turbine efficiency. This means a larger number of producing wells and a larger collection system. Since the turbine efficiency increase would be at best from 70% to as high as 85%, the improvement in system efficiency would be a net increase of 10 to 15% in power output. However, there would be temperature losses in the heat exchange process from the condensing steam to the heated working fluid. This is significant in the case of the low temperature geothermal sources. Thus there is some doubt that the use of a secondary working fluid would result in a net benefit once steam is produced. Thus the best system for producing electricity once flash steam has been obtained from the reservoir system is to use the steam directly in a turbine.

SECONDARY WORKING FLUID SYSTEMS APPLIED TO LIQUID-PRODUCING RESERVOIRS WITHOUT FLASHING

For the case of liquid-dominated fields in which the brine is available at the wellhead as a liquid, a secondary fluid working system can recover the thermal energy from the brine in a reversible fashion in a form in which the secondary fluid can be directly expanded through a turbine. A system for accomplishing this is shown in Figure 8.17. A hot, pressurized working fluid such as isobutane is expanded through a

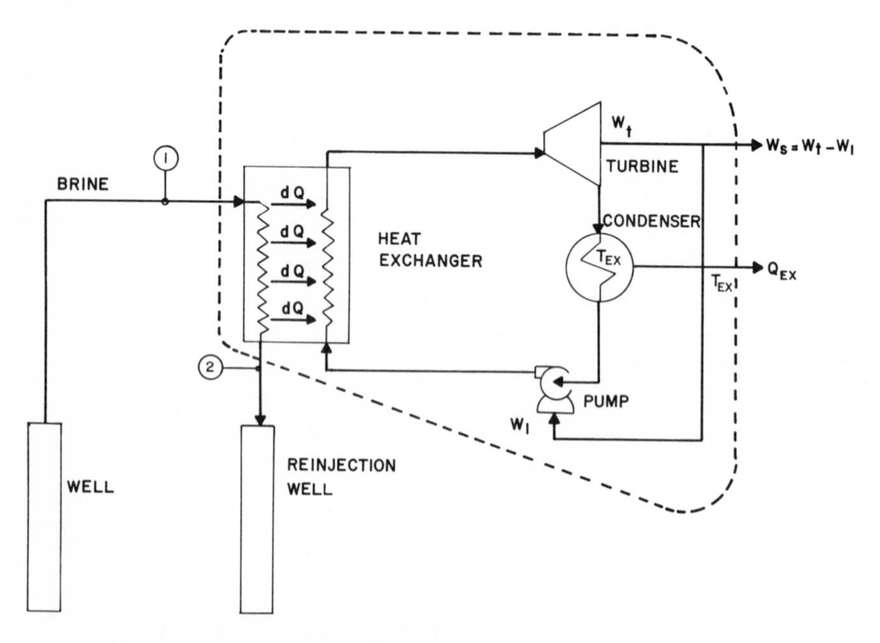

Figure 8.17 Process schematic for a binary cycle, that is, secondary working fluid, process plant for producing power from geothermal brine.

turbine, condensed, and pumped back up to the turbine inlet pressure for reheating in the heat exchanger. The key element of the system is a countercurrent heat exchanger in which the hot brine is transferring its heat to a working fluid with a very small temperature difference. This is possible if the cold working fluid entering the heat exchanger at the same end from which the cold brine is leaving is increased in temperature about the same amount as the brine decreases its temperature. This will maintain a small temperature difference between the two streams throughout the heat exchanger. This requires that the heat capacity of the working fluid and brine at the same locations throughout the heat exchanger remain in the same proportion.

Isobutane has been proposed as a convenient and economic working fluid for this system. The relationship between temperature and enthalpy for isobutane at various pressures is shown in Figure 8.18. Also shown in that figure is the temperature–enthalpy relationship of hot water and a 3 wt.% geothermal brine. The brine enthalpy–temperature relationship is linear reflecting its constant heat capacity. Note that for very high pressures the isobutane

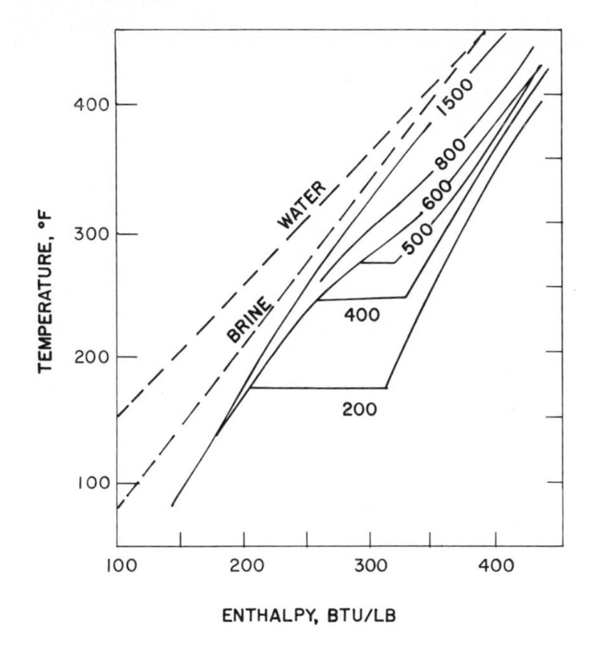

Figure 8.18 Temperature–enthalpy relationship for isobutane for various pressures. The temperature–enthalpy relationship of water and a brine containing 3 wt.% total dissolved solids is shown by the dashed lines.

enthalpy–temperature relationship is approximately linear. If the isobutane system is operated at this high pressure so that its heat capacity is nearly constant, then the countercurrent heat exchanger can be operated so as to maintain small temperature differences throughout its length. However, if the isobutane pressure is lower, say for example, 500 psia, then its heat capacity varies substantially over the range. In fact in the subcritical pressure region at its boiling point, the enthalpy change is very large for a fixed temperature. In this case, the heat exchanger might operate with small temperature differences in some parts of the heat exchanger but not others because of the nonlinear nature of the enthalpy–temperature relationship of isobutane.

If the heat exchanger is operating with small temperature differences and, therefore, nearly reversibly, then the work produced by the secondary system can be calculated using the basic system equation, Equation 8.1. This work will be a function of the temperatures and pressures of the brine entering and leaving the heat exchanger and of the exhaust, that is, condenser, temperature.

Applying Equation 8.1 to the system enclosed by the dashed line in Figure 8.17 where the temperature at which the heat is being rejected by the system is the condenser temperature, the expression for the reversible work of the secondary working fluid system is obtained:

$$W_{SR} = (h_1 - h_2) - T_{EX}(s_1 - s_2) \tag{8.14}$$

In this equation, the subscripts 1 and 2 refer to the points 1 and 2 in Figure 8.17. The net reversible work from the system is also equal to the turbine work less the pump work as indicated in Figure 8.17:

$$W_{SR} = W_{tR} - W_{1R} \tag{8.15}$$

The corresponding equation for the actual work produced is

$$W_{SA} = W_{tA} - W_{1A} \tag{8.16}$$

The actual turbine and pump work are related to the reversible work by the appropriate efficiencies:

$$W_{tA} = e_t W_{tR} \tag{8.17}$$

$$W_{1R} = e_p W_{1A} \tag{8.18}$$

Consequently, the actual and reversible works for the system can be related by combining Equations 8.14 through 8.18:

$$W_{SA} = e_t W_{SR} - W_{1R}\left(\frac{1}{e_1} - e_t\right) \tag{8.19a}$$

This assumes the heat effect of the pump is negligible, which is usually true. The latter term in Equation 8.19*a* will generally be small so that it can be neglected

$$W_{SA} \cong e_t W_{SR} \tag{8.19b}$$

Thus by combining Equations 8.14 and 8.19*b*, the actual work produced by a secondary fluid system with a condenser temperature T_{EX} and a reversible heat exchanger is

$$W_{SA} = e_t[h_1 - h_2 - T_{EX}(s_1 - s_2)] \tag{8.20}$$

Condensing Temperature

Note that, in general, the condenser temperature and the temperature of the cold brine rejected will be higher than the ambient conditions, so that the actual work obtained will be lowered by that effect as well as by turbine inefficiency below the work available from the brine as given by Equation 6.12. The amount of this loss is given by the difference between the work given by Equation 8.14 evaluated at the higher temperature and at ambient conditions thusly:

$$W_M - W_{SR} = h_2 - h_0 - T_0(s_1 - s_0) + T_{EX}(s_1 - s_2) \qquad (8.21)$$

The fractional work loss is then

$$\frac{(W_M - W_{SR})}{W_M} = \frac{[h_2 - h_0 - T_0(s_1 - s_0) + T_{EX}(s_1 - s_2)]}{[h_1 - h_0 - T_0(s_1 - s_0)]} \qquad (8.22)$$

This work loss amounts to about 1%/°F as shown by the plot of the equation for various brine inlet temperatures in Figure 8.19. It is sufficiently large so that consideration should be given to these temperature differences in designing and operating a particular plant. Another method of evaluating this effect is to use the differential of

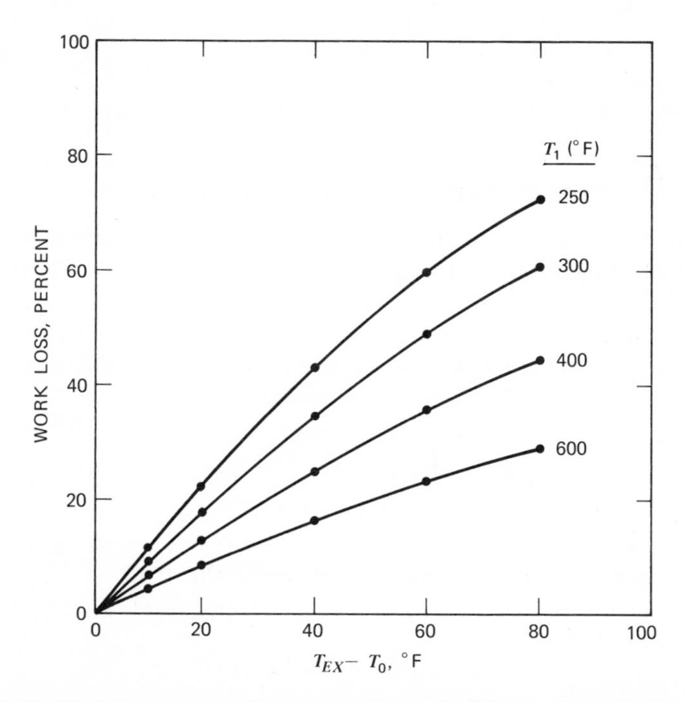

Figure 8.19 Work loss, expressed as percent, due to the difference between the ambient temperature and the condenser and brine rejection temperatures. The work loss is shown as a function of this temperature difference $T_{EX} - T_0$ for various brine inlet temperatures as indicated on the appropriate curves. It is assumed that heat exchanger is reversible so that the brine rejection temperature is equal to the condenser temperature. The brine properties are assumed to be those of water. Ambient temperature is 80°F.

the function W_M for evaluating small differences:

$$\frac{(W_M - W_{SR})}{W_M} = \left(\frac{dW_M}{dT_0}\right)\frac{(T_0 - T_{EX})}{W_M} \tag{8.23}$$

Since the heat capacity of water is nearly constant at a value close to unity, Equation 6.12 can be written as

$$W_M = c_p(T_1 - T_0) - T_0 c_p \ln\left(\frac{T_1}{T_0}\right) \tag{8.24}$$

Differentiating with respect to T_0 and inserting in Equation 8.23

$$\frac{(W_M - W_{RS})}{W_M} = \frac{(T_{EX} - T_0)[\ln(T_1/T_0)]}{[T_1 - T_0 - T_0 \ln(T_1/T_0)]} \tag{8.25a}$$

For brine temperatures of 300°F this is

$$\frac{(W_M - W_R)}{W_M} \approx (0.013)(T_{EX} - T_0) \tag{8.25b}$$

which is about 10% for an excess rejection temperature of about 10°F.

Heat Exchanger Temperature Difference

The importance of irreversible heat losses due to temperature differences in the heat exchange can be determined by calculating W_L the reversible work which is lost. One way to do this is to integrate the work output of infinitesimally small Carnot cycle systems operating over the temperature difference ΔT between the two fluids in the heat exchanger, Figure 8.7. From Equation 8.9

$$W_L = \int dW = \int_z \left(\frac{\Delta T}{T_z}\right) dQ_z \tag{8.26}$$

where z represents a particular location in the heat exchanger, dQ_z is the heat flow over a differential section of the heat exchanger per pound of brine flowing through the exchanger, and T_z is the brine temperature. Since

$$dQ_z = c_p\, dT_z \tag{8.27}$$

Equation 8.26 becomes

$$W_L = \int \left(\frac{c_p\, \Delta T}{T_z}\right) dT_z \tag{8.28}$$

Since the brine has a constant heat capacity and assuming that ΔT is constant throughout the heat exchanger, then

$$W_L = c_p \Delta T \ln \left(\frac{T_1}{T_2}\right) \tag{8.29}$$

Expressing this as a fraction of the reversible work output given by Equation 8.24 with T_{EX} or T_2 as is appropriate substituted for T_0,

$$\frac{W_L}{W_{SR}} = \frac{[\Delta T \ln (T_1/T_2)]}{[T_1 - T_2 - T_{EX} \ln (T_1/T_2)]} \tag{8.30a}$$

Alternatively, Equation 8.14 can be used for the reversible work output of the system resulting in

$$\frac{W_L}{W_{SR}} = \frac{[\Delta T \ln (T_1/T_2)]}{[h_1 - h_2 - T_{EX}(s_1 - s_2)]} \tag{8.30b}$$

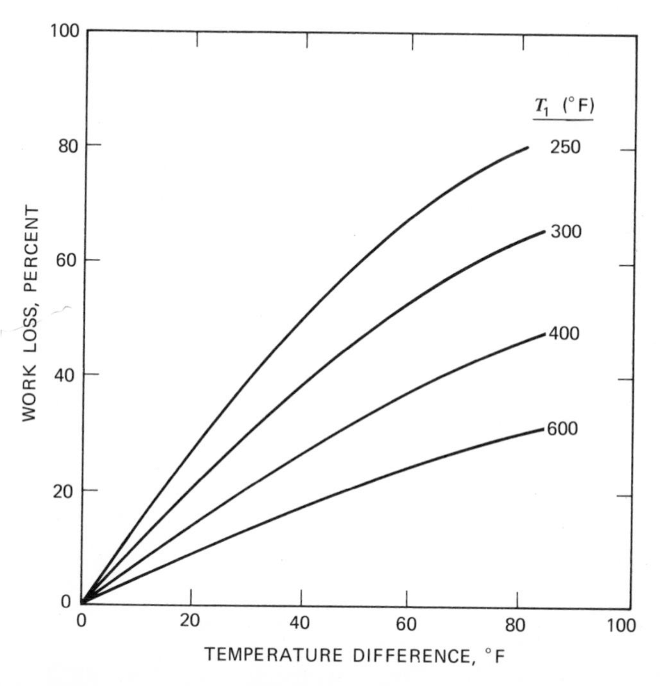

Figure 8.20 Work loss, Equation 10.30, due to a uniform finite temperature difference in the heat exchanger of a secondary fluid working system expressed as percent of the reversible work for various brine inlet temperatures T_1. The heat capacity of the brine is assumed equal to unity, and the exhaust temperature of the system is 130°F.

For a reasonable ΔT, say 50°F, for $T_2 = T_{EX} + \Delta T =$ 130 + 50 + 460°R, and for $T_1 = 760$°R, W_L/W_{SR} from Equation 8.30 has the value 0.6. Consequently, even if the same quantity of heat is transferred, the effect of irreversibility in the heat exchanger is quite significant. This fractional work loss, as given by Equation 8.30 and plotted in Figure 8.20, shows that the loss is significant unless the heat exchanger ΔT can be kept to small values. Note that Equations 8.30*a* and *b* and Figure 8.20 include both the effect of the heat exchanger inefficiency as given by Equation 8.29 and the loss of recoverable thermal energy. Less thermal energy is recoverable because the exhaust temperature of the brine T_2 is higher than the exhaust temperature T_{EX} of the system by the heat exchanger temperature difference ΔT.

Correcting Equation 8.19*a* for the losses due to finite temperature

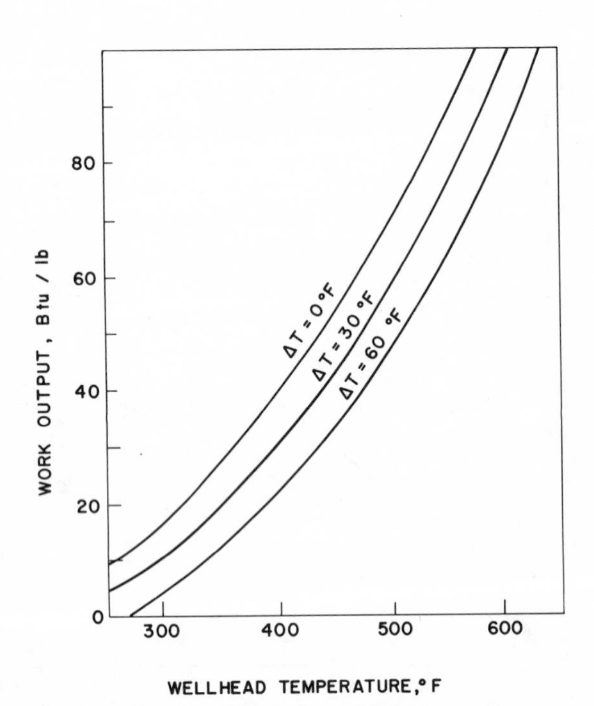

Figure 8.21 Work output from a secondary working fluid system with a conventional heat exchanger as a function of brine temperature for various assumed irreversible losses in the heat exchanger as indicated by the temperature difference on the graphs. The thermodynamic properties of water are assumed. The exhaust temperature has been taken as 130°F, turbine efficiency as 78%, and working fluid pump efficiency as 100%.

differences in the heat exchanger gives

$$W_{SA} = e_t\left[W_{SR} - c_p\,\Delta T \ln\left(\frac{T_1}{T_2}\right)\right] - W_{1R}\left(\frac{1}{e_1} - e_t\right) \tag{8.31}$$

where W_{SR} is given by Equation 8.14 as before. Neglecting the pump irreversibility and using Equation 8.14,

$$W_{SA} = e_t\left[h_1 - h_2 - T_{EX}(s_1 - s_2) - c_p\,\Delta T \ln\left(\frac{T_1}{T_2}\right)\right] \tag{8.32}$$

This equation is plotted in Figure 8.21 for various values of ΔT.

Conventional Heat Exchangers

The cost of heat exchangers for such a system are substantial relative to the total plant capital cost. For example, the installed cost of heat exchangers for a geothermal plant processing 10 million lb/hr of brine at 350°F to produce about 50 MW of power is $20 million. The cost of all power plant equipment other than the heat exchangers would be approximately the same as the power plant cost given in Table 8.2, about $8 million, because the same equipment would be necessary. Table 8.3 shows the total capital cost for such a system. The power plant cost and well transmission system cost in this table were computed using the information in Table 8.2 to give a total plant cost of $36 million. Of this total installed cost, the heat exchangers contribute 55%. The cost of the turbine might be somewhat less than shown because of the higher pressure of the system. On the other hand, there would be an additional cost for equipment necessary to supply makeup working fluid. The cost of the heat exchangers given in this table were computed assuming that they were constructed completely of carbon steel, that the temperature difference between the working fluid and the brine was 30°F and that the pressure on the isobutane side of the heat exchanger was 600 lb/in.^{2} gauge. Because the cost of the heat exchangers is so large relative to the total capital cost of the equipment, the best system design will be one in which the heat exchanger cost is minimized.

Direct Contact Heat Exchange

Because of the high rate of heat transfer expected in a direct contact heat exchanger, this type of heat exchanger has been proposed for use in geothermal power plants. One type of direct contact heat exchanger

Table 8.3 Capital Cost (1975 Dollars) of a Secondary Working Fluid Geothermal Power Plant Using Shell and Tube Heat Exchangers to Process 10,000,000 lb/hr of 350°F and Produce 50 MW of Electrical Power

Basis of Heat Exchanger Cost	
Area	$0.7 \times 10^6 ft^2$
Number of 5000 ft^2 Exchangers	142
Cost per Exchanger	$ 35,000
Equipment Cost, 142 Exchangers	$ 5,000,000
Installed Cost, 4 x Eqt. Cost	$ 20,000,000
System Capital Cost	
Power Plant Cost	
Heat Exchangers, Conventional	$ 20,000,000
All Other Costs	8,000,000
Total Power Plant Costs	$ 28,000,000
Well & Transmission System Cost	8,000,000
Total System Capital Cost	$ 36,000,000

is shown in Figure 8.22. In this type of heat exchanger, the hot brine enters the top of the column and flows down across a series of perforated plates. The cold isobutane enters the bottom of the column and because it has a lower density than the brine floats to the top of each section where it flows through the holes in the plates. These holes disperse the isobutane so that it passes through the brine in a finely divided state causing good heat transfer between the two fluids. Other types of direct contact heat exchange columns could be used.

This system may have the advantage that scale deposition problems of conventional heat exchangers are avoided. Such heat exchangers have not yet been operated using geothermal brines, and the importance of solids deposition on their performance is unknown. The installed cost of direct contact heat exchange columns for the power plant described in Table 8.3 is about $3 million as compared with the $20 million for conventional shell and tube heat exchangers. Thus this type of heat exchanger offers considerable advantage over conventional heat exchangers. In fact, the use of such a heat exchanger is probably necessary if a secondary working fluid system is to be

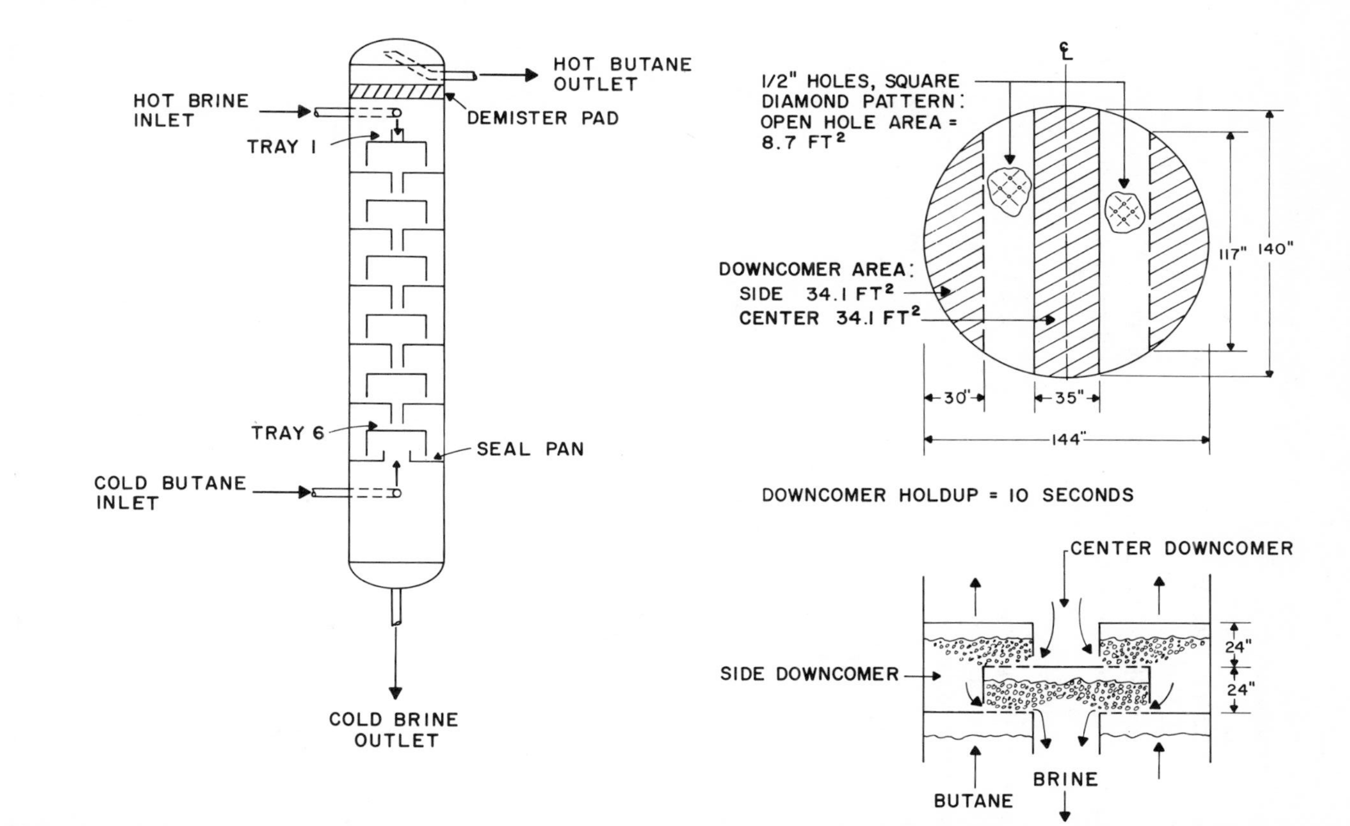

Figure 8.22 Schematic drawing of a direct contact heat exchange column. The internal parts of the column can be various designs such as sieve plates, bubble plates, or packing. This particular column has sieve plates.

economically viable for power generation from low temperature geothermal brine sources.

A system for generating power using a direct contact heat exchanger and a secondary working fluid such as isobutane is shown in Figure 8.23. In this system, the brine from the well is increased in pressure by a pump requiring an amount of work W_2 to raise it to the column operating pressure that might typically be 600 psia. The isobutane from the condenser is fed to the column by a pump requiring an amount of work W_1. In the direct contact heat exchange column that is operating in countercurrent flow, heat is transferred from the hot brine to the isobutane over a small temperature difference. The smaller the temperature difference, the greater the efficiency of the system will be. The cold brine leaving the column is still at high pressure, and an amount of work W_3 can be recovered from it before discharging it at atmospheric pressure. The hot isobutane is then expanded through a turbine to obtain the work output W_t. The net work output for the system is then equal to

$$W_{SA} = W_t - (W_1 + W_2 - W_3) \tag{8.33}$$

The turbine work is calculated from the turbine efficiency, and the reversible work for the turbine, as described in Chapter 7, according to the equation

$$W_t = e_t W_{Rt} \tag{8.34}$$

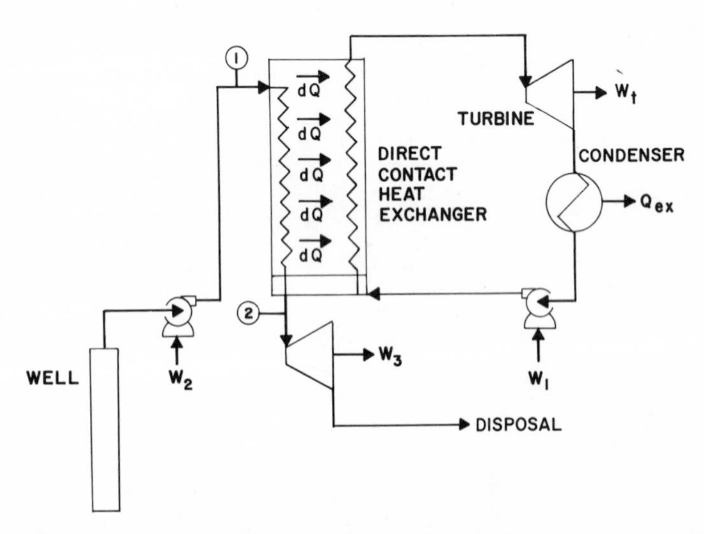

Figure 8.23 Process schematic for a direct contact heat exchange secondary fluid working system for power generation from geothermal brine.

The work of each of the liquid pumps or expanders is determined by multiplying the pressure volume work by the efficiencies of the device:

$$W_i = e_i V \Delta P \tag{8.35}$$

Combining these equations gives the work output. Alternatively, the work output can be calculated using Equation 8.31 applied to the points 1 and 2 in Figure 8.23 and deducting the net pump work required for the brine:

$$W_{SA} = e_t \left[h_1 - h_2 - T_{EX}(s_1 - s_2) - c_p \Delta T \ln\left(\frac{T_1}{T_2}\right) \right] - W_{1R}\left(\frac{1}{e_1} - e_t\right) - (W_2 - W_3) \tag{8.36}$$

This equation is plotted in Figure 8.24 as a function of brine temperature assuming that the properties of the brine are the same as those of water. Comparison of Figures 8.21 and 8.24 shows that the sum of all

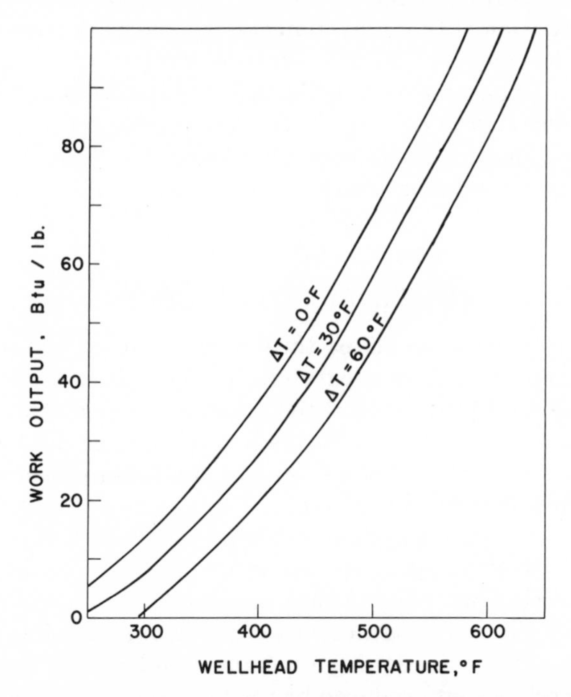

Figure 8.24 The work output from a direct contact heat exchange system assuming an exhaust temperature of 130°F, pump efficiency of 75%, and turbine efficiency of 78%. The solid line is Equation 8.36. The thermodynamic properties of water are assumed.

the pump work for a direct contact secondary working fluid system is not significant for wellhead temperatures above 350°F. At 350°F and $\Delta T = 30°F$, for example, the pump work terms will decrease the work output of a direct contact system 14% below that of a conventional heat exchange system with a 100% efficient working fluid pump. At 300°F, on the other hand, this loss is 29%, which is significant.

No direct contact heat exchanger will operate with zero temperature differences. There must be some finite temperature difference to cause heat transfer. A reasonable temperature difference is 10 to 30°F in such a column. In some portions of the column, there would be a smaller temperature difference and in other portions a larger temperature difference because of the nonlinear relationship between the enthalpy and temperature of isobutane as compared with the linear relationship for brine. For such a case, the computation of the work must be carried out for the secondary working fluid system by making an individual computation on the heat exchanger and the turbine. The performance of the heat exchanger can be computed assuming certain efficiency of operation of the column and by writing an energy balance around it and each of the individual stages. The work output from the turbine can be computed for the reversible expansion assuming either an isentropic expansion or a polytropic expansion, according to the procedures outlined in Chapter 7. The work output is then calculated according to Equations 8.33 through 8.35 for the direct contact heat exchanger shown schematically in Figure 8.23.

TWO-PHASE FLOW SYSTEMS

Another method for overcoming the difficulty caused by the necessity of a large number of flash separators for efficient exploitation of hot brine as described previously is the direct expansion of the hot brine through a machine that converts the hot pressurized fluid energy into mechanical rotating shaft work. Any of the expansion machines for two-phase flow described in Chapter 7 might be used.

The process for such a system, as shown in Figure 8.25, is relatively simple. It involves pumping the hot brine up to the required pressure for the expansion process and then expanding it to a sufficiently low pressure so the discharge temperature is as near as possible to ambient temperature. This will require a vacuum condenser.

The net work output produced by the two-phase expansion machine is the reversible work times the efficiency of the expansion machine, calculated as described in Chapter 7, less the pump work required to

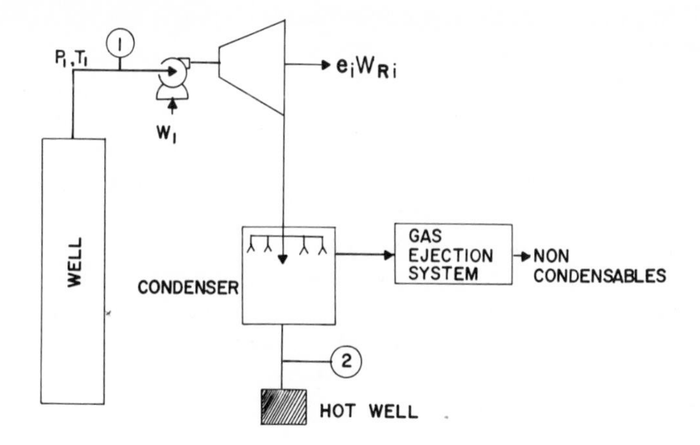

Figure 8.25 Process flow schematic for power generation from geothermal brines using a two-phase process system.

raise the brine pressure to that required for the expansion machine. Thus the net work output is

$$W_{PA} = e_i W_{Ri} - W_1 \tag{8.37}$$

where the subscript i refers to some unidentified expansion machine. Consequently, the net work output for such a system is strongly influenced by the efficiency of the two-phase expansion machine.

For a 100% efficient machine the net work output is given by the reversible work output for any process system, Equation 8.1, that applied to the conditions shown in Figure 8.25 is

$$W_{Ri} = (h_1 - h_2) - T_{EX}(s_1 - s_2) \tag{8.38}$$

The subscripts 1 and 2 in this equation apply to the points 1 and 2 in Figure 8.25. The pump work required for pressurizing the brine is equal to the reversible work divided by the efficiency of the pump:

$$W_1 = \frac{W_{1R}}{e_1} \tag{8.39}$$

where the subscript 1 refers to the brine pump. Combining Equations 8.38 and 8.39 with Equation 8.37 gives

$$W_{PA} = e_i[h_1 - h_2 - T_{EX}(s_1 - s_2)] - \frac{W_{1R}}{e_1} \tag{8.40}$$

This equation provides a simple method for estimating the net power output for a total flow system providing that the efficiencies of the expansion machine and pump are known.

The efficiency of some of these machines is described in Chapter 7 based on consideration of the mechanism of the expansion process. Since these machines have only been partially demonstrated, if at all, the efficiency of these systems must be considered unknown in the range given in Table 7.1. The net work output for a two-phase total flow expansion process as given by Equation 8.40 is shown in Figure 8.26 for various assumed expansion efficiencies.

Different two-phase expansion machines may be compared by their thermodynamic efficiencies. In some cases, such as the Bi-phase system described in Chapter 10, the expansion machine may be more complex than one simple process step and may involve several steps. Nevertheless, the comparison of that system with other systems is made on the basis of the efficiency of the expansion process between the inlet and outlet conditions. The capital cost and operating cost of the system must be taken into account and also be considered when

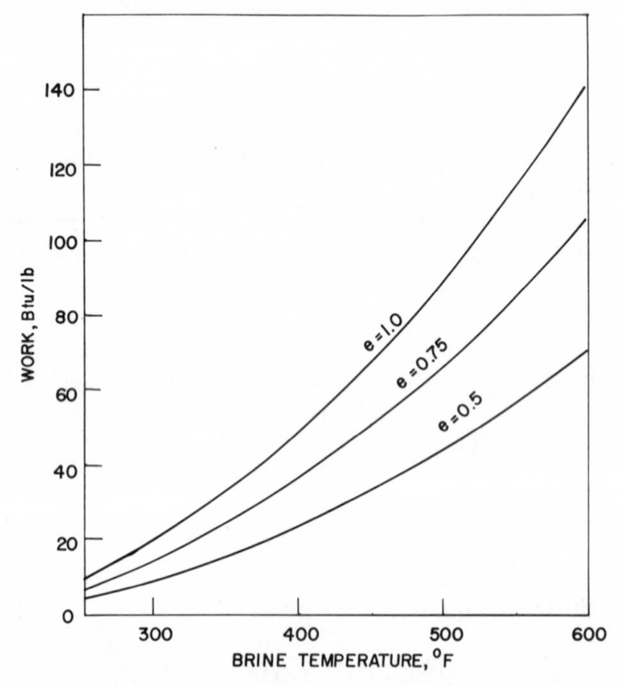

Figure 8.26 Work output for a two-phase expansion system for various assumed efficiencies of the expansion machine. The exhaust temperature is 130°F.

designing a two-phase expansion system and comparing it with other types of systems.

COMPARISON OF PROCESS SYSTEMS AND SUMMARY

The maximum work from a steam source, whether it is produced from flashed brine or from a steam-dominated field, is obtained by expanding the steam through a turbine. In this case, the net work output is equal to the work obtained from a reversible expansion of the steam times the turbine efficiency, which for low pressure geothermal turbines is 70 to 75%. For hot brine systems, the net system work for multiflash separator systems, a secondary working fluid system, and for total flow systems is shown in Figure 8.27. Also shown is the limiting work output for a perfectly reversible process as a function of the brine temperature and the assumed exhaust temperature. For this figure, two-phase expansion machine efficiencies of 50 and 80% were

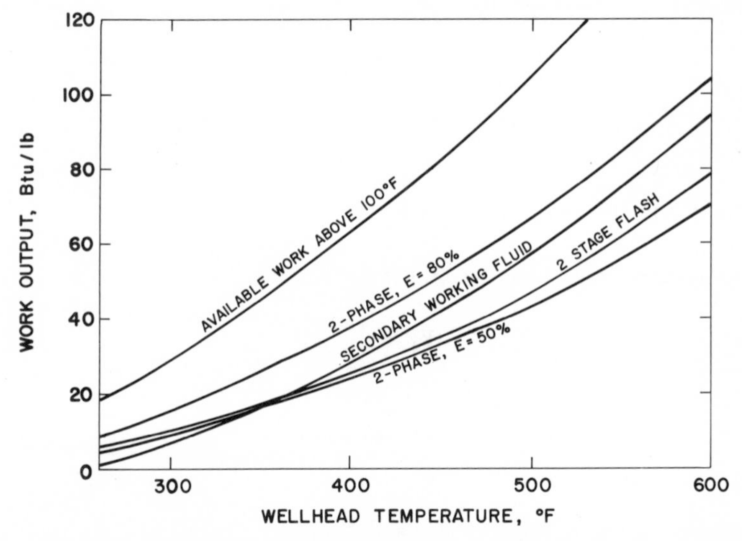

Figure 8.27 The work output of various systems as a function of brine temperature compared with the maximum work output possible from a reversible process. The multiflash separator system assumes a turbine efficiency of 70%. The secondary working fluid system assumes a turbine efficiency of 78% and a heat exchanger temperature difference of 30°F. The total flow system is shown for machine efficiencies of 50 and 80%. The maximum, that is, available, work output is for ambient and brine exhaust temperatures of 100°F, which is 30°F below the assumed condenser temperatures of the systems.

used to correspond to those described in Chapter 7. For the secondary working fluid system, the data for a reasonably efficient direct contact heat exchanger and a turbine with an efficiency of 78% is used. The multiflash system curve is given for a two-stage system, although a higher temperature reservoir would support a larger number of flash stages.

This comparison shows that all three of the foregoing systems produce a work output as a function of temperature that is similar. The secondary working fluid system has an advantage over the multiflash system because of the problem of irreversible temperature losses in the flash process. If the efficiency of a two-phase expansion process is 50%, then it does not compare well with the other systems. However, if an 80% efficiency can be achieved with a two-phase expansion system, then it compares advantageously with the other systems.

Representative equations derived for these systems are listed below for convenience.

Any system, reversible work

$$W_R = (h_1 - h_2) - T_{EX}(s_1 - s_2) \tag{8.1}$$

Steam dominated reservoir, direct expansion

$$W_A = e_t W_R$$

where W_R is the reversible work defined by Equation 8.1.

Multistage flash separator, steam turbine

$$W_A = \sum W_i \tag{8.13c}$$

where $W_i = e_{ti}(x_{i-1} - x_i)(h_{vi} - h_{EX})$.

The limiting work for an infinite number of flash stages is

$$W_A = e_t W_R$$

where W_R is the reversible work defined by Equation 8.1.

Secondary working fluid system

$$W_{SA} = e_t\left[W_R - c_p \Delta T \ln\left(\frac{T_1}{T_2}\right)\right] - W_{1R}\left(\frac{1}{e_1} - e_t\right) - (W_2 - W_3)$$

where W_R is the reversible work defined by Equation 8.1.

Two-phase expansion machine systems

$$W_{PA} = e_i W_R - \frac{W_{1R}}{e_1}$$

where W_R is the reversible work defined by Equation 8.1.

NOMENCLATURE

c_p	heat capacity, Btu/lb °F
e	efficiency, %/100
h	specific enthalpy, Btu/lb
H	Enthalpy, Btu
m	mass, lb
Q	heat, Btu/lb; Q_H, positive flowing in; Q_C, positive flowing out
Q_{EX}	exhaust heat per pound of brine processed, Btu/lb
s	specific entropy, Btu/lb °R
S	entropy, Btu/°R
T	absolute temperature, °R
x	mass ratio brine to brine feed, pounds of brine per pound of brine fed to process
W^*	work, Btu
W	specific work, Btu/lb
z	spatial location

Subscripts

A	actual work
C	cold
EX	exhaust (rejection) conditions from the system
H	hot
i	ith stage or ith power conversion device
L	liquid or lost work
M	reversible process discharging at ambient temperature and pressure
O	ambient conditions
p	pump
P	two-phase expansion
R	reversible
RC	cold reservoir
RH	hot reservoir

S secondary fluid or binary system
t turbine
V vapor
z spatial location

REFERENCES

General

Austin, A. L., "Prospects for Advances in Energy Conversion Technologies for Geothermal Energy Development," Lawrence Livermore Laboratory, Univ. California, Livermore, Calif., 1975.

Elliott, D. G., "Comparison of Brine Production Methods and Conversion Processes for Geothermal Electric Power Generation," Jet Propulsion Laboratory, Pasadena, Calif., 1975.

Walter, R. A., et al., "Thermodynamic Modeling of Geothermal Power Plants," Battelle Pacific Northwest Laboratories, May 1975.

Economics

Barr, R. C., "What is the Outlook for Geothermal Power?" *Oil Gas J.*, 148 (May 12, 1975).

Bloomster, C. H., "An Economic Model for Geothermal Cost Analysis," Battelle, Richland, Wash., 1975.

Bloomster, C. H., et al., "Geocost: A Computer Program for Geothermal Cost Analysis," Battelle, Richland, Wash., 1975.

Futures Group, "A Technology Assessment of Geothermal Energy Resource Development," NSF-RA-X-75-011, National Science Foundation, Washington, D.C., April 15, 1975.

Geothermal Energy Magazine, "Geysers Development Costs Table," *Geothermal Energy*, **2**(10), 56 (October 1974).

Greider, R., "Economic Considerations for Geothermal Exploration in the Western United States," *Bull. Geotherm. Res. Council*, **3**(3) (May 1974).

Holt, B., and J. Brugman, "Investment and Operating Costs of Binary Cycle Geothermal Power Plants," presentation to the NSF Conference on Research for the Development of Geothermal Energy Resources, Pasadena, Calif., 1974.

House, P. A., et al., "Potential Power Generation and Gas Production from Gulf Coast Geopressure Reservoirs," Lawrence Livermore Laboratory, Univ. Calif., Livermore, Calif., 1975.

Kunze, J. F., "What If the Water Isn't Hot Enough?" *Geotherm. Energy Mag.* (May 1975).

McCabe, B. C., "Practical Aspects of a Viable Geothermal Energy Program," presentation to the Second U.N. symposium on the Development and Use of Geothermal Resources, San Francisco, May 20–29, 1975.

Sapre, Z. R., and R. J. Schoeppel, "Technological and Economic Assessment of Electrical Power Generation from Geothermal Hot Water," presentation to the Second U.N. Symposium on the Development and Use of Geothermal Resources, San Francisco, Calif., May 20–29, 1975.

Towse, D., "The Economics of Geothermal Heat as an Alternate Fuel," presentation to the Fall Meeting of the Society of Mining Engineers, AIME, Salt Lake City, Utah, 1975.

Operating Plants

Aikawa, K., and M. Soda, "Advanced Design in Hatchobaru Geothermal Power Station," Second U.N. Symposium of the Development and Use of Geothermal Resources, San Francisco, May 1975.

Dal Secco, A., "Turbocompressors for Geothermal Plants," *Geothermics*, Special Issues 2, **2**, 819 (1970).

Dal Secco, A., "Geothermal Plants Gas Removal from Jet Condensers," Second U.N. Symposium on the Development and Utilization of Geothermal Resources, San Francisco, May 1975.

Guiza, J. L., "Power Generation at Cerro Prieto Geothermal Field," Second U.N. Symposium on the Development and Use of Geothermal Resources, San Francisco, May 1975.

Villa, F. P., "Geothermal Plants in Italy: Their Evolution and Problems," Second U.N. Symposium on the Development and Utilization of Geothermal Resources, San Francisco, May 1975.

Usui, T., and K. Aikawa, "Engineering and Design Features of the Otake Geothermal Power Plant," *Geothermics*, Special Issue 2, **2**, 1533 (1970).

Secondary working fluid systems

Cortez, D. H., B. Holt, and A. Hutchinson, "Advanced Binary Cycles for Geothermal Power Generation," Ben Holt Co., Pasadena, Calif., 1973.

Kilhara, D. H., and P. S. Fukunaga, "Working Fluid Selection and Preliminary Heat Exchanger Design for a Rankine Cycle Geothermal Power Plant," Dept. Mechanical Engineering, Univ. Hawaii, 1975.

Krishnamohan, K., et al., "Studies of Direct Contact Heat Transfer Between Immiscible Liquids in a Perforated Plate Tower, *Chem. Process Eng. (Bombay)*, 27–31 (July 1970).

Sheinbaum, I., "Direct Contact Heat Exchangers in Geothermal Power Production," A.S.M.E. Heat Transfer Conference, San Francisco, Calif., August 11–13, 1975.

CHAPTER 9

Thermal Utilization and Mineral Recovery

Thermal energy is the form of energy first produced in the conversion of coal, oil, nuclear, and geothermal resources to useful work. This energy is delivered and used in a variety of ways at a temperature level ranging from ambient conditions up to the highest temperatures that technology has been able to contain, generally 1000 to 3000°F. In the case of plasma systems, these maximum temperatures can be even higher. Geothermal energy delivered to the surface of the earth in the form of hot brine or steam is particularly useful for lower temperature applications.

REQUIREMENTS FOR THERMAL ENERGY IN THE UNITED STATES

The largest use of thermal energy in the United States, excluding conversion to electricity or direct mechanical drive as used for transportation is 12×10^{15} Btu/yr for space heating, 43% of the total thermal consumption for applications under 400°F. Water heating is the next largest single use in the United States and accounts for the consumption of 2.5×10^{15} Btu. These applications require thermal energy at a temperature sufficiently high to produce comfortable conditions or sufficiently hot water, probably about 120 to 150°F for space heating and about 180°F for water heating. In both cases, however, use of countercurrent heat exchange would enable the thermal energy source to be utilized down to temperatures below even 70°F in polar climates. Thus for utilization of geothermal energy in cases where the energy content of the material to be heated is relatively low, countercurrent heat exchange may significantly increase the amount of available energy that is transported and utilized.

Air conditioning and refrigeration are two other large users of energy in the United States. Thermally driven air conditioning systems have been used for many years and require that the heat be delivered at a temperature above 250°F. Refrigeration generally requires a somewhat higher temperature, about 350 to 400°F, because of the greater power requirement for cooling the refrigerated material to below freezing temperatures. The temperature level required for refrigeration or air conditioning machines depends primarily on whether the heat is rejected from the cooling machine to air or water. The use of water increases the efficiency of the machine and enables a lower temperature heat source to be used. Since most refrigeration and air conditioning machines are located in the vicinity of needs for hot water, it may be possible for an overall system properly designed to use the heat rejected from the refrigeration and air conditioning machines for part of the hot water heating requirements.

The food industry is another large user of thermal energy. The temperatures required by the food industry range from low temperatures for drying to high temperatures for cooking. The temperatures for blanching and drying are generally below 200°F so as not to degrade the food or feed products. On the other hand, cooking requires temperatures in the neighborhood of 400°F or higher.

Drying of industrial products is another large consumer of thermal energy, which is particularly applicable for geothermal energy sources. Many industrial drying processes, as with food, must be kept at low temperatures. For example, the drying of raw rubber must be done at a low temperature so as not to polymerize and degrade the uncured rubber. Drying of paper and lumber products are other examples of large users of thermal energy that must be done within a temperature range of 200 to 300°F.

Throughout the chemical industry, process heat in the form of steam is commonly used at temperatures ranging from 200 to 500°F or higher in exceptional cases. More than 80% of the steam used in petroleum refineries is delivered at a temperature below 350°F. The processing of minerals such as clay and cements are, in a sense, part of the chemical industry that are often considered separately. These industries also use thermal energy at 200°F and higher.

Thermal energy is used for producing electricity. This is the largest use of thermal energy in the United States and accounts for 45% of the total United States' energy consumption. Thermal energy for powering electrical production plants is generally delivered at a temperature of about 1000°F. This is the highest temperature that technology has been able to contain in the materials necessary for steam power plant machinery.

The other large use of thermal energy in the United States is the transportation industry. Transportation accounts for approximately 16% of the United States' energy consumption. Most transportation is driven by combustion engines in which the thermal energy of the combustion products is contained by cooled metallic surfaces and thus are able to withstand higher temperatures.

The uses of thermal energy, the temperature at which it is required, and the United States' annual energy consumption in each of the categories discussed previously is summarized in Table 9.1. This table shows that 30% of the United States' energy consumption is for space and process heat below 400°F, much of which might therefore be supplied by geothermal reservoirs.

TRANSPORT OF THERMAL ENERGY

All of the previously described processes require that thermal energy be provided at a particular location. The thermal energy may be provided at that location by transport of the fuel to the specific point of application, such as in internal combustion engines, or it may involve conversion of the thermal energy into a hot fluid stream, which is then transported to the site of application, such as steam circulation in processing plants. Thermal energy is also transported as steam in some cities in the United States, although the application in this manner is less common today than in the past.

In general, much of the United States' economy is based on the transport of energy in the form of electricity. In some cases, however, it may be more advantageous to transport energy in its thermal form rather than as electricity. Since geothermal energy is already in the thermal form, this chapter will deal with the transport of thermal

Table 9.1 Thermal Energy Consumption in the United States for Various Temperature Ranges and End Uses

Temperature (°F)	Use	Consumption[a] (10^{15} Btu/yr)	(10^3 MW)
100	Greenhouses	2.0	66
to	Space Heating	12.0	401
150	Water Heating	2.5	84
to	Drying	1.0	33
210	Blanching/Cooking	1.2	40
to	Multiple Effect Evaporation	0.4	13
250	Industrial Processing	0.8	27
to	Air Conditioning	1.8	60
300	Industrial Processing	2.0	67
to	Refrigeration	1.5	50
350			
to	Industrial Processing	2.5	84
400			
<400	Subtotal, Heating Below 400°F	28	925
>1000	For Electricity Production	42	1404
>1000	High Temperature Processes	8	267
>1000	Direct Drive-Transportation	15	502
	Total	93	3098

Source. Reistad, 1975; Towse, 1975

[a] Energy is usually expressed in Btu for thermal energy and megawatt-hours for electrical energy. Since Chapters 9 and 10 discuss both types of energy, it is convenient to express both energy forms in the same units, arbitrarily megawatts hereafter, for the purpose of comparison and interpretation. For example, the 1.4×10^6 MW of thermal energy consumed for electricity production produces 0.45×10^6 MW of electricity assuming 33% conversion efficiency.

energy. No consideration will be given here to the transport of energy in other forms.

TRANSPORT OF GEOTHERMAL ENERGY

Present practice of geothermal energy utilization requires that the heat stored in the earth's crust be transported from hydrothermal systems as sensible heat of brine water or steam to the surface of the earth. Generally this is accomplished by allowing the geothermal brine to flow to the surface under its own pressure through a well. Usually the output of more than one well is required to power a given generating station, for distribution to a district heating system, or for powering an industrial plant. Consequently, the geothermal brine is collected in a central location from a set of wells. This typically requires a transport distance of several thousand feet. The collected fluid is then processed, either to make electricity or to use directly as process or space heat. In the latter case, the thermal energy must be transported and then distributed to the consumers.

The thermal power P_{th} being transported by F lb/hr brine with enthalpy h is

$$P_{th} = F(h - h_{EX}) \tag{9.1a}$$

A useful relation, obtained by a simple transformation of units, for the quantity of geothermal brine in thousands of pounds per second required to transport megawatts* of thermal power is

$$F^* = \frac{0.95\, P_{th}^*}{(h - h_{EX})} \tag{9.1b}$$

For brine with a heat capacity of 0.95 Btu/lb°F, such as 4% salt water, this becomes

$$F^* = \frac{P_{th}^*}{(T - T_{EX})} \tag{9.1c}$$

Thus the amount of brine to be transported in thousands of pounds per second is equal to the amount of heat required, in megawatts, divided by the sensible heat stored in the brine. The sensible heat is approximately equal to the usable temperature drop of the brine. This makes a convenient and easily remembered conversion between brine flow rate and transported heat in megawatts. This relation also emphasizes the fact that a very large quantity of fluid must be transported for a

*See footnote for Table 9.1.

commercial-sized electrical power plant. For example a 50-MW electrical power plant operating at 12% efficiency and utilizing brine with a usable temperature drop of 200°F would require 2100 lb/s, which is 280 gal/s or 1 million gal/hr, or brine. Assuming each well can deliver 1000 gal/min of brine, then 17 wells would be required for the 50-MW electrical generating station.

The transport of geothermal-produced thermal energy in a fluid stream involves the design of a system that balances cost of the capital equipment, the distance of transport, heat losses, and operating costs. To understand the significance of these various factors in system design, it is important to have some idea of the value of the product being transported.

VALUE OF GEOTHERMAL ENERGY

The value of the product obtained from geothermal energy depends on the end use. This is shown by a comparison of the use of geothermal energy for the production of electricity, of industrial process heat, and of residential space and hot water heating. This comparison will be based on the fuel value, that is, cost for useful output. Fossil fuels are delivered to electrical power plants and industrial processing plants in large quantities at a price of approximately $2.50/1,000,000 Btu. The same fossil fuels delivered to residences or small commercial establishments is about $4.00/1,000,000 Btu. The conversion efficiency for electric plants using fossil fuels is 37% of the available thermal energy. Industrial process plant boilers have efficiencies that at best might be 85%. The losses of thermal heat in the boilers are due mainly to the thermal energy carried out with the exhaust gases, called stack losses. On the other hand, residential and small commercial establishments utilize only 70% of the thermal energy available in the combustion products. A comparison of fossil fuels for these various applications requires a correction for the appropriate conversion efficiency.

Since geothermal energy is normally delivered for thermal purposes as hot water, it can be utilized with near-100% efficiency for space or hot water heating of residential or small commercial establishments. This assumes that the energy is measured above the lowest temperature that can be utilized. Consequently, a higher price can be paid for delivered geothermal energy than for the equivalent Btu content of a fossil fuel. In fact, one could afford to pay up to a ratio of 100% divided by 70% more for the geothermal energy than for the fossil fuel energy.

Table 9.2 Effect of End Use on the Value of Geothermal Fluids. The Delivered Value Is That Value That Gives the Same Output as an Equivalent Value of Fossil Fuel

	Fossil Fuels		Geothermal Fluid	
End Use	Price ($\$/10^6$ Btu)	Conversion Efficiency (%)	Conversion Efficiency (%)	Delivered Value ($\$/10^6$ Btu)
Electricity	2.50	37	5	0.34
Industrial Processes	2.50	85	100	2.94
Residential & Small Commerical	4.00	70	100	5.71

As shown in Table 9.2 this amounts to $5.71. Similarly, geothermal energy that is used for process purposes would be more valuable than the fossil fuels in the ratio of 100 to 85%, In the case of utilization of geothermal energy for making electricity, however, the situation is just reversed because geothermal energy can be converted to electricity with an efficiency of only 1 to 10%. With 5% as a typical conversion efficiency, geothermal energy would be of value in the ratio of 5 to 37%, or about $0.34.

Thus geothermal energy used for heating purposes is about 20 times as valuable as it would be if it were used for generating electricity. This means that one could afford 20 times the investment for transmission of the geothermal energy used for thermal purposes over the investment for electricity generation. For example, the transmission of geothermal fluids for making electricity becomes uneconomic at distances much greater than 0.5 to 1 mi. On the other hand, operating district heating systems in Iceland, described later, demonstrate that thermal energy can be transported more than 10 mi and still produce an economically viable system. Waste hot water from the Lardarello geothermal power plant is transported as far as 66 mi to provide space heating (Haseler, 1975).

FLUID TRANSPORT CONSIDERATIONS FOR COLLECTION SYSTEMS

Most of the special flow considerations arise because a large quantity of geothermal fluid in the form of a gas, a liquid, or a two-phase mixture of the two is being transported. Additional difficulties arise because this fluid may contain a large quantity of dissolved salts. Some special considerations are:

1. In the reservoir and piping system consideration must be given to the region where phase change is occurring because deposition of salts can cause plugging of the reservoir strata, the well casing, or the piping. This means that careful attention must be given to the pressure drop and/or the method of pressurizing the fluid.

2. At high flow rates in the well, flashing will occur resulting in a phase change from liquid to gas. During the transition, the flow may be a liquid-in-gas phase or it may be slugs of gas separated by liquid. It also might be a gas core with a liquid annulus. Associated with each of these types of flows will be special problems. For example:

   a. High velocity flow of a gas stream with entrained liquid droplets will cause erosion of impacted surfaces. Two pieces of equipment where this can be expected to occur are the separator and turbine blades. Otherwise two-phase flow is not a serious problem and can be used successfully with appropriate design.

   b. Surging may occur in the well due to alternating plugs of gas and liquid.

3. If the brine flashes to form steam, it may be necessary to separate the hot brine from the steam in a separator. Design of the separator requires consideration of the droplet size distribution, the centrifugal forces applied, and the drag forces of the gas on the liquid droplets.

4. Special problems may also arise because of the high temperature of the brine:

   a. The piping system used to transport the fluid will be subjected to significant thermal expansions. This includes both the well casing and the surface piping used to transport the fluid to the central processing plant. Distances are typically 0.25 mi or more, and expansion loops or devices will be necessary in the piping system. No special considerations are usually given to the temperature problems associated with expansion of the well casing, even though there are significant stresses associated with the temperature change of the casing.

b. The piping system must be insulated to prevent heat losses.

c. Packing and seals of pumps and valves must be given special consideration to prevent leakage that in turn might cause corrosion, erosion, and/or mineral deposition.

5. Because of the large volume of gases handled there will be a significant noise problem associated with release of the steam from flash separators to the atmosphere or with operation of cooling towers.

6. Because of the presence of 0.2 to 2% by volume of noncondensable gases in flashed steam, gas extraction systems are necessary for operating vacuum condensers.

DESIGN CONSIDERATIONS FOR SENSIBLE HEAT TRANSPORT

The amount of available thermal power transported by a pipe system is the sensible heat in the fluid fed to the system less any heat losses. The heat loss can be determined by integrating the loss, as shown schematically in Figure 9.1, over the entire length of the pipe. Assuming that all the resistance to heat transfer is through the insulation, then the heat loss is given by (McAdams, 1954)

$$dq = Fc_p dT = \left\{\frac{2\pi k}{\ln[(D + D_x)/D]}\right\}(T - T_0)\,dl \tag{9.2}$$

Rearranging and integrating from the inlet 1 to the point of delivery 2

$$Fc_p \ln\left[\frac{(T_1 - T_0)}{(T_2 - T_0)}\right] = \left\{\frac{2\pi k}{\ln[D + D_x)/D]}\right\} L \tag{9.3}$$

A solution more convenient for subsequent use is obtained when the temperature loss $T_1 - T_2$ over the length of pipe is small compared with the temperature difference $T - T_0$. For this case, $T - T_0$ can be assumed constant and the heat loss integrated directly to give

$$q_L = a_1(T - T_0)L \tag{9.4}$$

where $a_1 = 2\pi k/\ln[(D + D_x)/D]$. Thus the net delivered thermal power P_{Dth} is, using Equations 9.1 and 9.4,

$$P_{Dth} = F(h - h_{EX}) - a_1(T - T_0)L \tag{9.5}$$

If transport of thermal energy is to be efficient, the heat losses $a_1(T_1 - T_0)L$ should be small compared with the total power available for transport, $F(h - h_{EX})$. If water is the transporting fluid and the exhaust temperature is equal to ambient temperature, the latter term

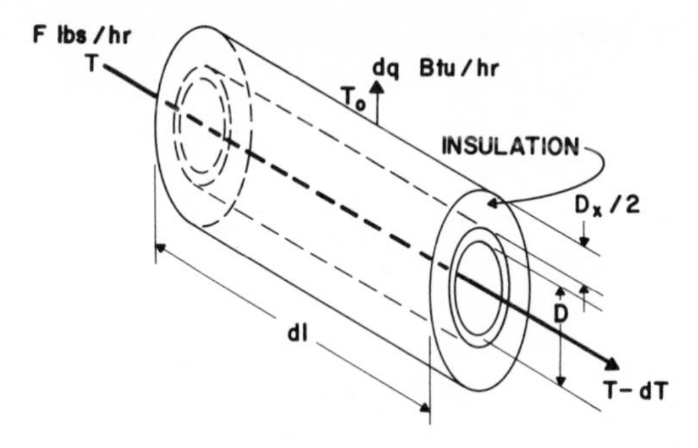

Figure 9.1 Heat loss $c_p dT$ from a hot fluid flowing at F lb/hr through an insulated pipe. Heat losses dq to the colder surroundings from the pipe of length dl must equal the heat loss from the fluid $Fc_p dT$.

becomes

$$P_{th} = F(T - T_0) \tag{9.6}$$

Dividing the heat losses, Equation 9.4, by Equation 9.6 gives the ratio of power loss to power available from the geothermal wells by the pipeline system:

$$\frac{q_L}{P_{th}} = \frac{a_1 L}{F} \tag{9.7}$$

For insulation thickness equal to one-third the pipe diameter, a_1 has the value 1 so that by rearranging Equation 9.7

$$\frac{L}{F} = \frac{q_L}{P_{th}} \tag{9.8a}$$

For the losses to be less than 10% of the power, Equation 9.8a states that the pipeline length in feet should be less than one-tenth the flow rate in pounds per hour. Furthermore, the distance over which the thermal energy can be transported economically is directly proportional to the power being transported. Because of the significance of this simple relationship to the utilization of geothermal energy, it is shown graphically in Figure 9.2 for various fractions of power loss. The equation shown in that figure is obtained from Equation 9.8a by changing units and using Equation 9.1c so that

$$L^* = \left(\frac{q_L}{P_{th}}\right)\left[\frac{6470 P_{th}^*}{(T - T_{EX})}\right] \tag{9.8b}$$

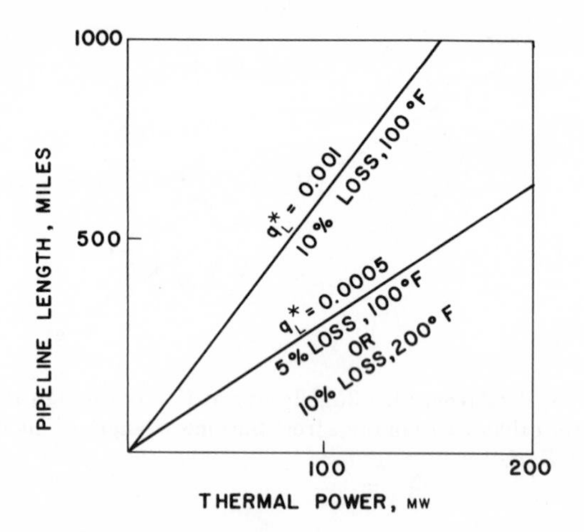

Figure 9.2 Pipeline distance as a function of the power delivered to the pipeline for various ratios of q_L*. Combinations of efficiency and temperature of the transport medium above its useful temperature are shown that correspond to the given value of q_L*. The computation applies to a fluid with a heat capacity of unity, an insulation with $k = 0.1$, and a thickness equal to one-third the pipe diameter.

where L^* is in miles. Rewriting

$$L^* = 6470 q_L^* P_{th}^* \tag{9.8c}$$

where $q_L^* = q_L / P_{th}(T - T_{EX})$. This figure demonstrates that thermal energy at sufficiently large power levels can be transported large distances with reasonable thermal efficiency.

The two dominant elements of cost for transporting geothermal brine or a secondary heat transmission fluid are the capital and maintenance costs for the pipeline and the pumping costs. The cost of capital and maintenance can be estimated as a fixed proportion of the total installed cost of the pipeline and the pumping system. Thus the annual cost C_{an} for 8760 hr/yr is given by

$$C_{an} = 8760(A_c C_L + A_c C_p W_p + C_e W_p) \tag{9.9}$$

where A_c is the fractional charge on capital due to maintenance and cost of capital per hour; C_L is the installed cost of pipeline, \$; C_p is the installed cost of pumps, \$/(Btu/hr); C_e is the cost of electrical energy, \$/Btu; and W_p is the pumping power, Btu/hr.

The installed cost C_L of a pipe system for a particular range of pipe diameters and transport distances is proportional to the length of installed pipe L and the pipe diameter D. Thus

$$C_L = L C_L^* D \tag{9.10}$$

where C_L^* is the proportionality constant. It is the cost of the installed pipe per foot length per foot of diameter. The cost of an installed insulated underground pipe system in the size range of interest for geothermal applications involving megawatts of power for distances of several or more miles is shown in Table 9.3. From this table the value of C_L^* is \$166/ft^2 for a single pipe installation and \$292/ft^2 for a system requiring a return pipe. Due to specific terrain requirements and quality of installation, this number will vary, but these figures are a good first estimate.

The cost of pumping is the cost of electrical energy times the pump work corrected for inefficiencies. The pump work is the volume of fluid transported times the pressure drop through the pipe divided by the pump efficiency. The efficiency of centrifugal pumps operating near their design point is 75 to 80%. Since the rate of volume transport is the mass flow rate F divided by the density ρ, the pump power is given by

$$W_P = \left(\frac{F}{\rho}\right)\left(\frac{\Delta p}{e_p J}\right) \tag{9.11}$$

The pressure drop through the pipe is given by

$$\Delta P = \frac{32 f L F^2}{\pi^2 \rho D^5 g_c} \tag{9.12}$$

Table 9.3 Typical Cost of Installed Insulated Underground Pipe Systems

	Cost for 6 Inch Pipe System ($/ft)			
	Single Pipe System		With Return Pipe	
Materials				
6 in. pipe, 2 in. insulation, and 25 yr. galvanized duct	$40		$70	
Fittings @ 30%	12		21	
Total		$52		$91
Installation @ 60% of Materials		31		55
Installed Cost, Total		$83		$146
Cost per foot per ft of Diameter, ($/ft^2)		$166		$292

Combining Equations 9.11 and 9.12 gives the pumping power:

$$W_P = \left(\frac{32f}{\pi^2\rho^2 e_p J g_c}\right)\left(\frac{LF^3}{D^5}\right) \tag{9.13}$$

The installed cost of a large centrifugal pump delivering about 10^6 Btu/hr or 500 HP is \$0.12/(Btu/hr) of pump power. A geothermal transmission system will require many large pumps, and consequently the installed cost of the pumps for the system will be equal to $C_p W_p$ where C_p is 0.12. Thus the total cost C_t per unit pump work of the pumps and pumping will be

$$C_t = A_c C_p + C_e \tag{9.14}$$

The cost per unit of thermal power C^* as a function of the pipe diameter is obtained by combining Equations 9.10, 9.13, and 9.14 with Equation 9.9 and dividing the result by Equation 9.5 after converting to a common time base. The result is

$$C^* = \frac{(A_c L C_L^* D + C_t a_2 L F^3/D^5)}{[F(h - h_{EX}) - a_1(T_1 - T_0)L]} \tag{9.15}$$

where

$$a_2 = \frac{32f}{\pi^2\rho^2 e_p J g_c} \tag{9.16}$$

For a typical friction factor of 0.01, a_2 has the value of 5.2×10^{-17}.

Equation 9.15 for cost per unit power can be simplified by assuming the heat loss is small compared to transported power:

$$C^* = \frac{(A_c L C_L^* D + C_t a_2 L F^3/D^5)}{F(h - h_{EX})} \tag{9.17}$$

The pipe diameter that minimizes the cost per unit power C^* is obtained by taking the derivative of Equation 9.17, setting it equal to zero, and solving for the diameter. The result is

$$D = \left(\frac{a_2 C_t F^3}{5 A_c C_L^*}\right)^{1/6} \tag{9.18}$$

Using this diameter, the cost equation becomes

$$C^* = \frac{4.59 L (a_2 C_t)^{1/6} (A_c C_L^*)^{5/6}}{F^{1/2}(h - h_{EX})} \tag{9.19}$$

This equation shows that the cost is inversely proportional to the sensible heat $h - h_{EX}$ and inversely proportional to the square root of the flow rate. It also shows that the cost is approximately proportional to the annual cost of the installed system. Since little can be done about the delivery temperature of the brine from the geothermal energy source, the only variables that can be used to significantly reduce the annual cost are the sensible heat stored per pound of transported fluid and the annual cost of the installed pipe system. The first item can be improved by using a higher heat capacity fluid or one that undergoes a chemical reaction or phase change. The latter term can be affected by changing the financial structure to gain a lower annual charge per unit of capital and by decreasing the cost of the installed pipe. If transport of geothermal fluid were to become a significant element of the United States' economy, then the associated large-scale manufacture and installation of pipe systems would be expected to reduce the cost both because of volume and improvements in technology.

COLLECTION, DISTRIBUTION, AND TRANSMISSION IN THE REYKJAVIK, ICELAND SYSTEM

The Reykjavik Municipal District Heating System, Table 9.4, is designed to deliver a peak load of 350 MW to the city. The estimated

Table 9.4 Pump and Piping Requirements for Reykjavik Municipal District Heating System

Heat Load, Maximum	350 MW
Heat Load, Density	300 MW/mi^2
Collecting Mains	13 miles
Supply Mains	29 miles
Street Mains	91 miles
Pumps, 44 Wells	2926 hp
Pumps, 4 Main Plants	4750 hp
Pumps, 10 District Stations	4750 hp
Pumps, Total Installed	12426 hp
Pump Load at Capacity	9.2 MW
Pump Load as % of Output	2.6 %

Source. Zoega, 1974.

installed cost of the system is $50 million. The system involves collection of the well brine from 44 wells and its delivery through four main pumping plants and 10 district pumping plants by way of 13 mi of transmission piping and 120 mi of distribution piping. The collecting and supply mains are 6 in., and the 91 mi of street mains are 3 in. The total installed pump capacity for the system is 12,426 HP, which, if operating at full capacity, would use 9.2 MW or 2.6% of the maximum delivered power of the system. The system supplies 11,000 houses and 88,000 residents (Zoega, 1974).

The brine is pumped from bore holes with deep well pumps located at a depth of 300 to 400 ft. The geothermal brine from the bore holes is pumped through collecting mains to a central distributing plant. In these plants, the water is mixed with return water from the town to provide a supply temperature of 176°F. The water is distributed throughout the system at 176°F by one of two methods. One is the pumping of water through a single pipe system and its discharge to waste after passing through the system. In the second method, a two pipe system is used in which the cooled water is collected and returned to the central distributing plants. There it is mixed with the hot water from the reservoir to produce a distribution fluid at 176°F. Storage tanks are used to meet short duration peak loads. Oil-fired heating plants are used to meet peak load demands of cold weather.

The water in the Reykjavik system is low in mineral content. As a result of the low mineral content, the scaling and corrosion problems have not been severe. The newer high temperature geothermal fields being developed in Iceland, however, have higher mineral content, and scale deposition problems have been encountered. One method of handling these problems is to dilute the brine with fresh water. Another method, which is the proposed method for the Svartsengi field, is to use the flashed steam from the geothermal fluid to heat fresh water for circulation in the district heating system. For Svartsengi, this will be accomplished by collecting the steam at wellhead separators and transmitting it to a central plant where some power conversion will be achieved. The electrical power so produced will be used for pump work. It will amount to about 10% of the total thermal capacity of the system. The hot brine from the wellhead separators is to be transmitted separately to the central plant where it in turn will be flashed to produce steam and cooler water. This steam will be used in combination with the exhaust steam from the turbine to heat fresh water, which will be circulated through the district heating system. The planned thermal power of the system is 80 MW. A summary of the space heating systems in operation and planned for the near future in Iceland is presented in Table 9.5.

Table 9.5 Geothermal Space Heating Systems in Iceland

Location	Power (MW)	Number of People Served
Systems in Operation 1975		
Reykjavik	385	88,000
Seltjarnarnes	22	2,500
Selfoss	12	2,700
Hveragerdi	9	1,000
Hvammstangi	1.7	400
Saudarkrokur	9.4	1,800
Olafsfjordur	3.2	1,100
Dalvik	3.5	1,100
Hrisey	1.3	300
Husavik	15	2,100
Myvatn	1.1	200
Laugarvatn	1.5	200
42 Rural Schools	10	
Subtotal	474.7	101,400
Systems in Construction to be Commissioned 1975/76		
Kopavogur	46	11,700
Hafnarfjordur	43	10,900
Gardahreppur	14	3,400
Subtotal	103	26,000
Systems Being Planned		
Svartsengi Project	80	10,300
Subtotal	80	10,300
Total	657.7	137,200

Source. Einarsson, 1975.

ECONOMICS

The cost of delivered thermal energy from a geothermal reservoir consists of the costs of the well, the cost of the collection system, the cost of the piping and transmission system, and the cost of the distribution system if any. In addition, there will be a processing plant cost at the central collection point and pumping station that will vary depending on the salinity and temperature of the brine. In Iceland, for example, salinity problems are overcome by diluting the geothermal brine with enough fresh water that the solubilities of the salts are not exceeded during the cooling of the brine as it is used by the consumer. The costs for these elements have been well established by the systems that have been in operation in Iceland for 10 to 20 yr. Table 9.6 is a summary of these various costs. It is noted that the production costs for low temperature water are substantial, whereas for water at 300°F, the production costs are considerably lower. This is because the enthalpy content of the water, and consequently the power output of the well, is significantly higher as a result of the increased temperature of the well water. The transmission costs through the transmission pipe system and the pumping stations is a relatively small part of the cost. The distribution cost is the most significant part of the total cost of the delivered thermal energy. Consequently, if thermal heat is used for an industrial complex where the distribution costs would be minimized, the economics would be substantially improved.

The cost of the well varies with the depth as shown in Table 9.7. The cost of the piping system or the collection of the geothermal brine from

Table 9.6 Cost of Delivered Distributed Thermal Energy for the Reykjavik District Heating System in $ /MWh

	Wellhead Temperature, (°F)		
Item of Cost	212	248	302
Well and collection system	2.1	1.6	0.5
Transmission per mile	0.19-0.37	0.16-0.27	0.14-0.24
Distribution	3.8	3.8	3.8

Source. Einnarsson, 1975.

Table 9.7 Cost of a Geothermal Well for Various Well Depths and the Collection System Per Well

	Well Depth, (ft)			
	2000	4000	6000	8000
Cost of well, ($)	200,000	280,000	400,000	560,000
Cost of collection system ($)	30,000	30,000	30,000	30,000
Annual cost of well ($/yr)	60,000	84,000	120,000	168,000
Annual cost of collection system ($/yr)	3,000	3,000	3,000	3,000
Annual cost of both per well ($/yr)	63,000	87,000	123,000	171,000

the wellhead to a central location is typically $30,000/well. To be consistent with the costs as reported for electrical plants previously discussed and summarized in Table 8.2, the annual cost for operation of the well and collection system shown on Table 9.7 is computed as follows. The annual costs for operating the well consists of 20% of the well capital cost to cover replacement and effluent disposal plus 10% for fixed charges giving a total of 30% of the well cost for the annual charges. The annual charge for the collection system consists of fixed charges at 10%. Thus the annual cost of the well and collection system is 30% of the capital cost of the well plus 10% of the capital cost of the collection system. The cost for a given quantity of brine will depend on the flow rate produced from the well. The relationship between the well depth and well production rate on the cost of the well and collection system per million pounds of brine produced is shown in Table 9.8. Examination of this table shows that the cost of a typical well, 2×10^5 to 4×10^5 lb/hr and 4×10^3 to 6×10^3 ft deep, and collection system is roughly $20 to 70/1,000,000 lb of brine produced.

The cost of the transmission system can be calculated according to Equation 9.17 or 9.19. Using 180 Btu/lb for $h - h_{EX}$, a friction factor of 0.01, an electricity cost of $11.7/million Btu, a single pipe system $A_c = 2 \times 10^{-5}$ and a 350-MW system, Equation 9.19 gives

$$C^* = 2.6 \times 10^{-11} L \qquad (9.20)$$

Experience in Iceland summarized in Table 9.6 shows that the transmission cost is between $1.40 and $2.40/mi MWh of thermal power delivered for geothermal brine with wellhead temperatures of about

Table 9.8 Cost of the Well and Collection System in Dollars Per Million Pounds of Brine Produced

Flow Rate per Well (lb/hr)	Well Depth, (ft) 2000	4000	6000	8000
50,000	143	176	280	377
100,000	72	89	140	188
200,000	36	44	70	94
300,000	24	30	47	63
400,000	18	21	35	47
500,000	15	17	28	68

300°F. Using a value of $2/mi MWh for the cost of transmission agrees in general with Equation 9.20. The cost of the Reykjavik system is higher than given by Equation 9.20 because that system uses more than one transmission pipeline, and the pipe diameters are larger than optimum.

The cost of delivered thermal energy is then calculated by combining these two costs. The cost of the well and collection system per unit of thermal power is the cost of the well and collection system per 1,000,000 lb of brine divided by the thermal content of 1,000,000 lb of brine. The thermal content increases as the wellhead temperature increases. Consequently, the costs of the well and collection system per unit of power decrease with increasing temperature of the reservoir. This relationship is shown in Table 9.9, assuming a well and collection system cost of $40/million pounds of brine, a typical value as shown in Table 9.8. To this cost must be added the cost of the transmission system. Using Equation 9.20 for the cost of the transmission system gives the relationship presented in Table 9.9 for the cost of the transmission system for different wellhead temperatures. For convenience, the cost is given as a function of miles of installed transmission system. Adding these two costs gives the total delivered cost of the thermal energy to the point of distribution. If all of the thermal energy is utilized in a large complex at this point, then this is the net cost of the delivered thermal energy. This would be the case, for example, if the energy were being delivered to a large processing plant.

Table 9.9 Cost of Delivered Thermal Energy above 120°F Produced from Geothermal Reservoirs Assuming the Thermodynamic Properties of Water for the Brine for a 350 MW Thermal Energy System

	Cost in ($/MW-hr) for Wellhead Temperature of Brine, (°F)			
Item	200	300	400	500
Well and collection system[a]	1.4	0.7	0.5	0.33
Transmission, incl. pumping[b]	1.1L*	0.5L*	0.3L*	0.2L*
Delivered cost, not distributed	1.4+1.1L*	0.7+0.5L*	0.5+0.3L*	0.35+0.2L*
Distribution costs, if any[c]	3.8	3.8	no data	no data
Distributed delivered cost	5.2+1.1L*	4.5+0.5L*	--	--

[a]Well and collection system cost taken as $40 per million pounds of brine.

[b]L* is total length of pipe in miles.

[c]From Table 9.6.

If, however, the thermal energy is being distributed to a city district heating system, then the distribution costs must be added to the total delivered cost of the thermal energy to the point of distribution. Experience in Iceland as summarized by Table 9.6 shows that the distribution costs are $3.8/MWh. Adding this to the cost of the delivered thermal energy gives the total cost for distributed delivered thermal energy as a function of the reservoir temperature. Note that Table 9.9 is based on a system delivering 350 MW of thermal energy. If the size of a plant is considerably different from this then the transmission cost given in this table will not be valid.

Since the value of delivered thermal energy as shown in Table 9.2 is $5.71/million Btu or $19.48/MWh, then the number of miles that geothermal energy can be delivered for an equivalent cost can be calculated from the data in Table 9.9. The $19.48/MWh value is based on distributed delivered cost so that the corresponding cost must be used from Table 9.9. Thus for a reservoir with a wellhead temperature of 300°F,

$$4.5 + 0.5L^* = 19.48 \tag{9.21}$$

The solution to this equation states that thermal energy from a geothermal resource of 300°F can be transported a distance of 30 mi and distributed and still remain competitive with fossil fuels. Note that if the amount of power that is being transmitted is 3500 MW instead of 350 MW, then the optimum transmission costs according to Equation 9.19 will be reduced by the square root of the flow rate ratio or a factor of 3. This suggests that the delivery of thermal energy to a residential community 10 times the size of Reykjavik, Iceland, can be transmitted a distance of 90 mi and still remain competitive with fossil fuels. Note that this amount of power would require the collection of thermal energy from some 200 geothermal wells.

UTILIZATION OF GEOTHERMAL ENERGY FOR PROCESS HEAT

Geothermal energy is presently being used in New Zealand to provide drying of lumber at Tasmann. It is also being used at Rotoraru, New Zealand, for air conditioning a hotel. In Iceland, geothermal energy is being used for drying at a diatomaceous earth plant. Approximately 50 MW of geothermal energy is used for greenhouse heating in Iceland. The use of geothermal energy for heating of greenhouses is quite extensive and is used in most countries of the world that are in cold climates and have geothermal energy available. In addition, geothermal energy is used for the heating of swimming pools.

The use of geothermal energy for process heat is in its infant stages. In Iceland, for example, plans are underway for developing a seawater chemical complex based on geothermal energy. The geothermal energy would be used principally for evaporation purposes. A schematic diagram of the process is shown in Figure 9.3. This process is typical of the use of geothermal energy for mineral recovery. The details of production of salt, potassium chloride, calcium chloride, and bromine from the geothermal brine-powered mineral recovery process are discussed in the section on mineral production. Another example of the use of geothermal brine for process use is the recovery of sulfur by the use of hot geothermal brine to dissolve sulfur in the ground and pump it to the surface.

As the supply of oil and coal energy resources becomes more dear, the use of geothermal energy for process heat receives more attention from industrial corporations. Some corporations are actively studying geothermal energy resources with respect to the supply of energy for their process plants. Thus considerable development of the use of geothermal energy for process heat can be anticipated in the coming decades.

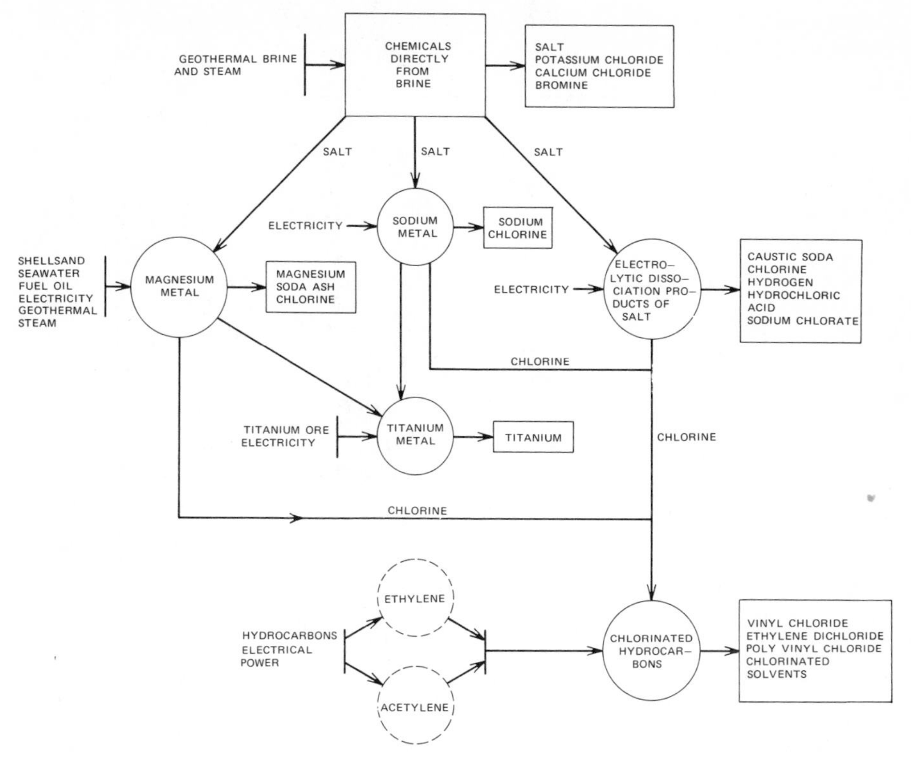

Figure 9.3 Seawater chemical complex proposed for Reykjanes Peninsula, Iceland (Lindal, 1975).

MINERAL RECOVERY UTILIZING GEOTHERMAL ENERGY

Whether the geothermal energy is to be used to recover minerals from the geothermal brine itself, from seawater, or from some other brine, the process is similar. Figure 9.4 shows a process schematic for the proposed recovery of salts from seawater or geothermal brine as part of the Reykjannes chemicals complex mentioned earlier.

The major constituents of brine are generally recovered by crystallization as a result of concentrating the brine. Sometimes this concentration is in conjunction with chemical effects such as pH change or addition of other salts. This is commonly called "salting out." The concentration process is done either by solar ponds or by multiple-effect evaporators. Examples of both approaches are found in current practice today. Minor constituents may be extracted by

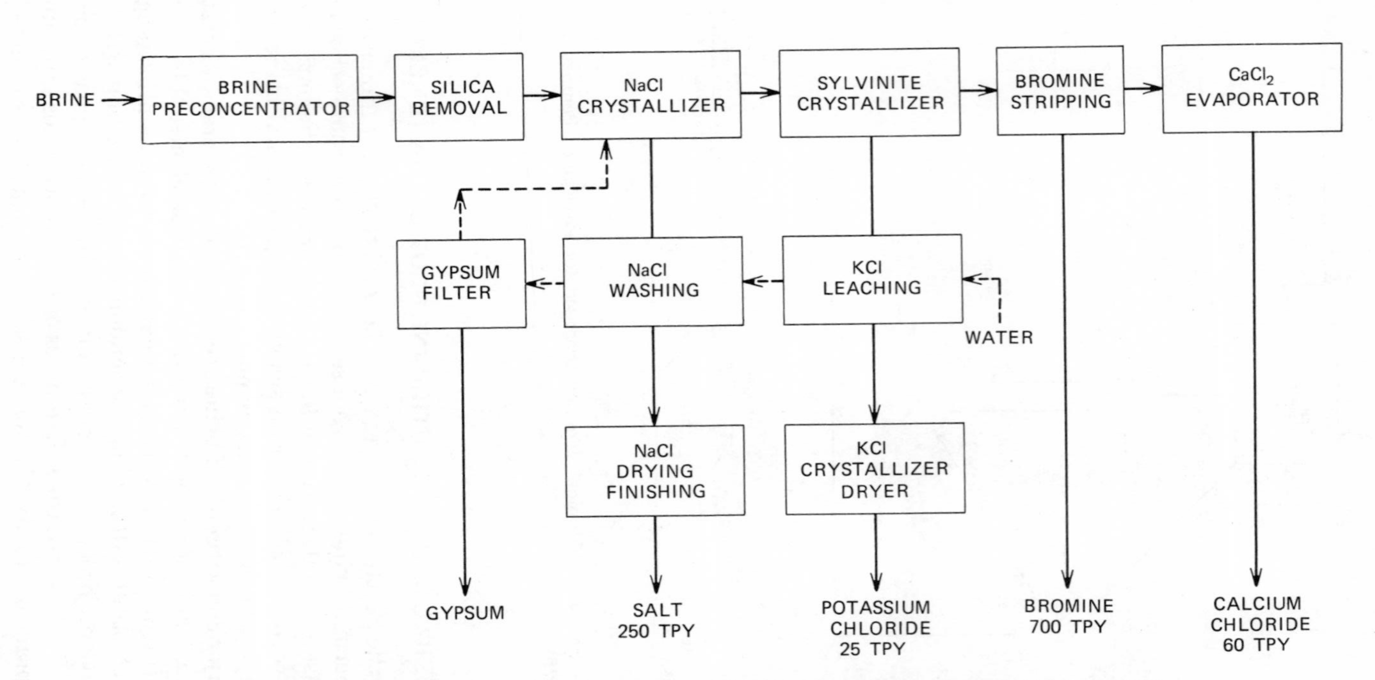

Figure 9.4 Recovery of minerals as proposed for the Reykjanes salt plant (Lindal, 1975).

crystallization, precipitation, or ion exchange, in particular liquid ion exchange.

Crystallization for Recovery of Major Constituents

The first step in the recovery of minerals from brines is usually the recovery of the more abundant species. In the case of geothermal brines, these are usually sodium, potassium, calcium, and chloride ions, which would be separated or recovered from the brine by crystallization. Such crystallization would be brought about by changing the temperature, by changing the concentration by evaporation, or by salting out. Usually the salting out is accomplished by addition of an ionic species that would cause one or more of the constituents to be precipitated in a controlled manner.

The processes used to remove the particular components and the purity in which they will be crystallized can be theoretically predicted by determining the solubility of the various mineral species that can be crystallized as a function of the concentration of the various ionic species. This means a knowledge of the solubility of the various forms of sodium chloride, potassium chloride, and calcium chloride in the presence of varying concentrations of their ionic components. A knowledge of the effect of the lesser abundant species must be superimposed on this.

The first step of the process is to cool the brine and partially evaporate it to precipitate sodium chloride. Evaporation is continued until the solubility limit of potassium chloride is reached. Then the solution is passed on to the next evaporation stage where the concentration is further increased to remove potassium chloride. When the solubility limit of the calcium chloride is reached, the solution is passed on to the next evaporation stage for removal of calcium chloride. Once these major elements have been removed, the solution then passes on to other process steps for removing the more valuable lower concentration elements. In each of the evaporation steps, the crystals that formed will not be pure because of the high concentration of the other species. Determination of precisely what crystal will form from the saturated solution requires a careful study of the phase chemistry and very likely some experimental studies.

Removal or recovery of the constituents present in lesser concentrations such as lithium, manganese, and/or magnesium might be accomplished by crystallization as previously described, by solvent extraction, or by adsorption on a carrier, as for example an ion exchange resin or activated carbon. Since crystallization of desirable species existing in dilute concentrations would necessitate concentrating the brine to a considerable extent, this approach may not be as economically attractive as ion exchange solvent extraction described later.

Ion Exchange Solvent Extraction

Liquid ion exchange such as the extraction of selected minerals by complexing with such organic reagents as carboxylic acid or amines will selectively remove certain metallic ions such as lithium into the organic phase. This organic phase may be further processed to enrich the desired component. Once this has been accomplished, the organic phase is stripped of the metallic component and recycled to the process. Considerable development of this technology took place with the recovery of uranium and related radioactive materials. One of the first applications of this technology to lower-cost products was the recovery of boron. Many examples of the application of this technology are found in current industrial practice today. Extraction by this technique can be accomplished with or without the prior removal of the major constituents; however, it will be technologically easier if the major constituents have been removed.

Prediction of the ion exchange solvent extraction processes for recovery of certain ionic species is accomplished by a knowledge of the solubility, or more correctly the distribution, of the ionic species between water and the ion exchange solvent. This information is necessary for estimating the amount of material recovered by the solvent or ion exchange media under various process conditions and estimating the cost of such a process. The performance of an adsorption material for recovery of certain species depends on the same thermodynamic properties; however, the data available in the literature is in limited experimental form and is not easily estimated from basic thermodynamic properties.

Reykjanes Geothermal Salt Plant

The process schematic for the geothermal salt plant portion of the sea chemicals complex for Iceland is shown in Figure 9.4. This is an example of the production of salts by evaporation. In this process, the brine is concentrated to first remove the silica. Since the solubility of silica is 100 ppm at a temperature of 100°F, the silica is readily removed prior to crystallizing other salts. After the silica is removed, the brine passes on to the sodium chloride crystallizer. The next step is the evaporation of the brine to produce the sodium chloride in that crystallizer. When the solubility of potassium chloride is reached, the brine passes on to the sylvinite crystallizer in which potassium chloride is formed. Water is used to wash the salts from each of the crystallizers to improve the concentration of the desired salt. A further step in this process, as a result of using seawater, is the production of bromine. In a final step, calcium chloride can be produced by evaporation of water below its saturation concentration.

Mineral Recovery from Concentrated Brines

An example of the application of the foregoing principles to the recovery of minerals from a concentrated brine is described later using the Salton Sea geothermal brine. A typical composition of this brine is shown in Tables 2.3 and 2.4. A hypothetical process including the various mineral recovery steps is shown schematically in Figure 9.5. Since the geothermal brines are saturated with respect to certain constituents, the flashing of water from the brine and cooling of the brine will result in precipitation and crystallization of certain constituents as shown by the first step in the process. Silica will almost always be saturated and so will precipitate. As described in Chapter 4, cations such as magnesium, strontium, iron, and aluminum, if present, will coprecipitate with the silica as a hydroxide or as a silicate. In the case of certain very concentrated Salton Sea brines and other geothermal brines, the solution may be saturated with respect to sodium chloride so that some will crystallize on cooling of the brine. Also, flashing of the brine with associated removal of dissolved gases will result in an increase in the pH of the brine and so decrease the solubility of certain constituents.

If the recovery process operates at atmospheric pressure, the maximum temperature of the brine would be about 220°F. The addition of calcium hydroxide will result in the precipitation of the insoluble hydroxides. After removal of the heavy metal hydroxides, an organic ion exchange solvent can be mixed with the hot brine solution to complex with certain select constituents. The separation of the organic phase and subsequent recovery of the selected minerals from the organic phase will result in the production of valuable minor constituents. Because the solubility of barium and strontium sulfate is much less than that of calcium sulfate as shown in Table 9.10, barium and strontium can be selectively precipitated even in the presence of higher concentrations of calcium as is typical in geothermal brines. Thus the addition of sulfate to the brine will result in the precipitation

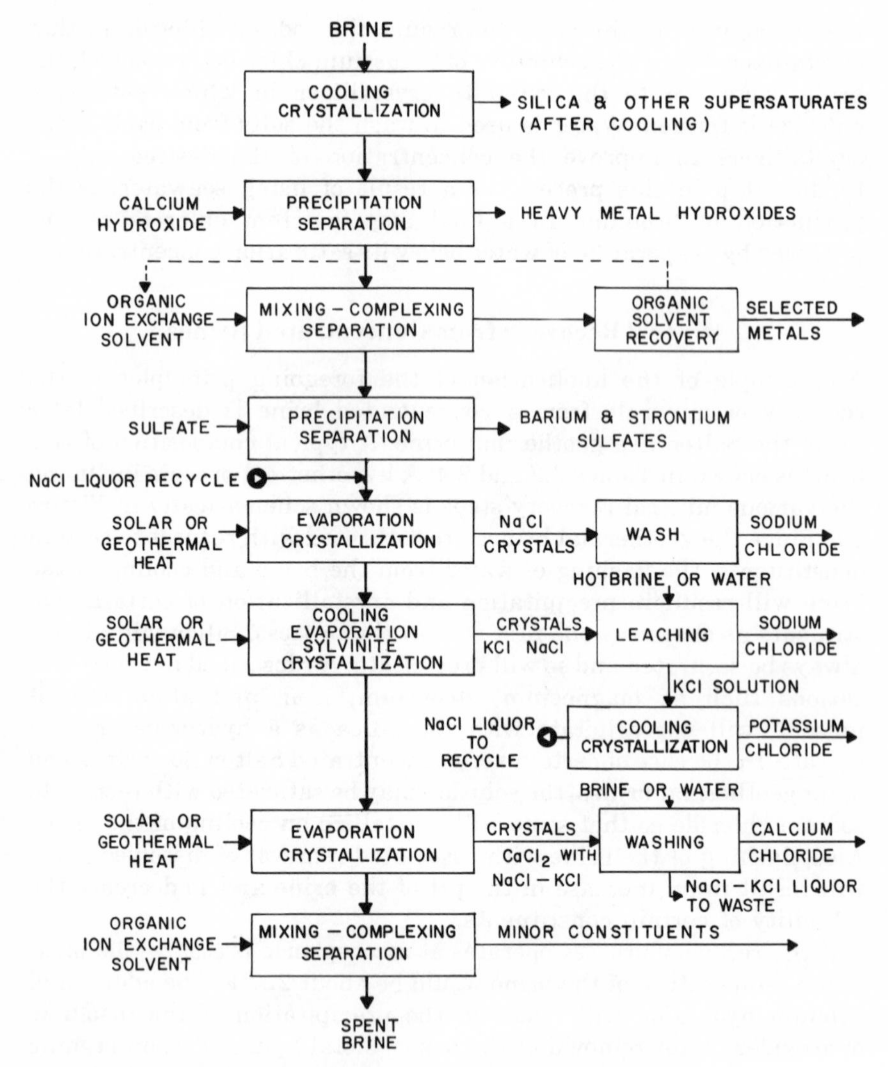

Figure 9.5 Schematic of a hypothetical process for recovery of minerals from a concentrated brine to exemplify process recovery techniques.

of barium and strontium sulfates, if they are present in sufficient quantity, thus recovering these metals from the brine as shown schematically, in Figure 9.5.

The next step in processing the brine is to remove the major constituents, namely sodium, potassium, and calcium chlorides. The sodium chloride is first removed by evaporating the solution until the solubility limit of the potassium–sodium chloride double salt, namely sylvinite, is reached. The evaporation of the brine for production of sodium chloride at 200°F rather than a lower temperature results in a greater recovery of sodium chloride and a liquor containing a higher concentration of potassium to sodium chloride, because the solubility of the sylvinite is greater at the higher temperature. Once the concentration has reached the point of crystallization of sylvinite, the brine solution can be passed from the sodium chloride crystallizer to the sylvinite crystallizer. At this point the cooling of the solution will result in the precipitation of potassium chloride. Further cooling and evaporation will result in the precipitation of a mixture of sodium and potassium chloride. These two steps could be carried out in separate crystallizers or in one crystallizer as shown in the schematic, Figure 9.5. The crystals of sodium and potassium chloride may be as high as 60% sodium chloride. These crystals are then leached with hot brine or with water from another source to dissolve the potassium chloride and produce a concentrated potassium chloride solution. Cooling of this hot potassium chloride solution results in the crystallization of potassium chloride in a sodium chloride liquor. This sodium chloride liquor can then be recycled to the process for the production of sodium chloride. The brine from the sylvinite crystallizer is a concentrated calcium chloride solution. Further evaporation results in the crystallization of calcium chloride with some sodium and potassium chloride salts. These crystals can be washed with brine or water to remove the sodium and potassium chlorides and produce an enriched calcium chloride product.

After the removal of the bulk constituents, the use of organic ion

Table 9.10 Solubility Product of Sulfates at 25°C

Substance	Solubility Product
Calcium sulfate	2.5×10^{-5}
Strontium sulfate	3.8×10^{-7}
Lead sulfate	1.0×10^{-8}
Barium sulfate	1.1×10^{-10}

Source. Weast, 1975, p. B-236.

exchange solvents and other chemical techniques for recovery of the minor and more valuable constituents is more easily accomplished. Only one step is shown in the schematic in Figure 9.5, but there could be a sequence of steps involving different types of ion exchange solvents, different pH of the solution, and other chemical reactants for recovery of certain of the minor constituents.

A technique for using the hot brine to exploit the difference in the temperature–solubility relationships of sodium chloride and sylvinite for the production of an enriched potassium chloride product is shown in the process schematic Figure 9.6. Most of the brine is processed in the manner described here and then passed on to the sylvinite evaporator. Here the potassium and sodium chloride is crystallized. The product from this crystallizer, which will contain about 60% sodium chloride, is leached with hot brine to remove the sodium chloride and produce a concentrated hot liquor of potassium chloride. The leaching operation is conducted at a temperature as close to the

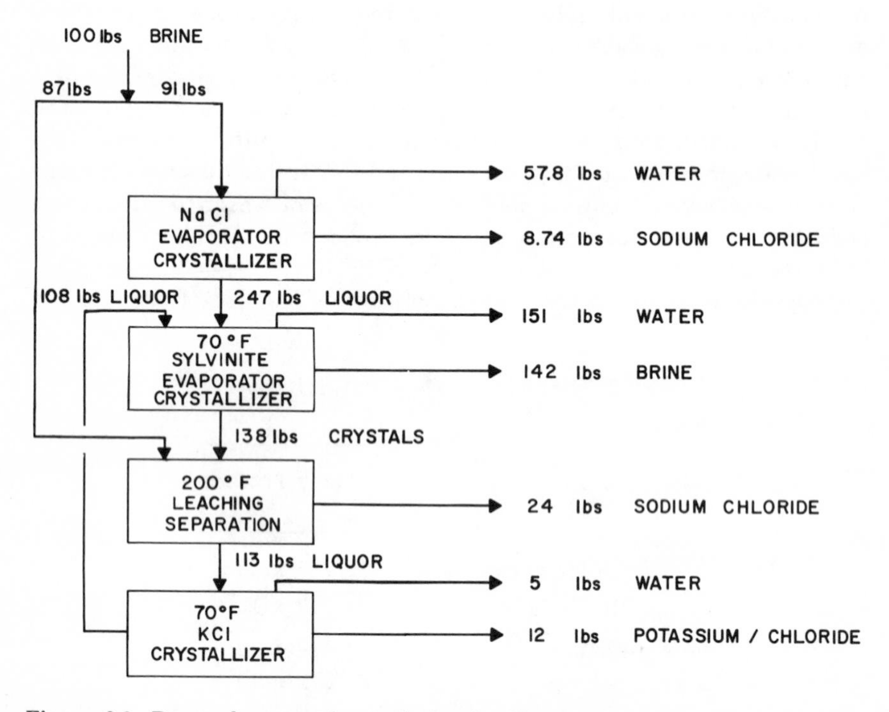

Figure 9.6 Proposed process for producing fertilizer grade, that is, 79% potassium chloride from a Salton Sea field geothermal brine (Berthold, 1975). Material balance flow rates are on the basis of 100 lb of brine fed to the process.

Greek Symbols

ρ density, lb/ft^3

Subscripts

an	annual
Dth	delivered thermal
e	electrical
EX	exhaust
L	pipeline
0	ambient conditions
p	pump
t	total
th	thermal
x	insulation

REFERENCES

Transmission

James, Russell, "The Presence of Steam Can Seriously Effect Hot Water Flow Measurements Utilizing an Orifice Meter," Second U.N. Symposium on the Development and Use of Geothermal Resources, San Francisco, May 20–29, 1975.

James, Russell, "Control Orifices Replace Steam Traps on Overland Transmission Pipelines," Geothermal Circular RJ 15, Department of Scientific and Industrial Research, Wairakei, New Zealand, 1975.

Haseler, A. E., "Direct Heating: An Annotated Bibliography," Property Services Agency Library, Lambreth Bridge House, London, 1975.

Lengquist, R., and Hansen, A., "Geothermal Steam Piping at Big Geysers, California, U.S.A., 1960–1975," Second U.N. Symposium on the Development and Use of Geothermal Resources, San Francisco, May 20–29, 1975.

McAdams, W. H., *Heat Transmission*, McGraw-Hill, New York, 1954.

Soda, Masahiro, et al., "Experimental Study on Transient Phenomena of Steam–Water Mixtures Flowing Through a Large Pipeline for Geothermal Power Stations," Second U.N. Symposium on the Development and Use of Geothermal Resources, San Francisco, May 20–29, 1975.

Takahashi, Y., et al., "An Experiment on Pipeline Transportation of Steam–Water Mixtures at Otake Geothermal Field," *Geothermics*, Special Issue 2, **2**(1), 882–891 (1970).

Utilization, Thermal

Arnorsson, S., et al., "Exploitation of Saline High-Temperature Water for Space Heating," Second U.N. Symposium on the Development and Use of Geothermal Resources, San Francisco, May 20–29, 1975.

Barnea, Joseph, "Economics of Multi-purpose Use of Geothermal Resources," Conference on Multi-purpose Use of Geothermal Energy, Klamath Falls, Oregon, October 7–9, 1974.

Burrows, W., "Utilization of Geothermal Energy in Rotorua, New Zealand," Conference on Multi-purpose Use of Geothermal Energy, Klamath Falls, Oregon, October 7–9, 1974.

Cooke, W. L., "Some Methods of dealing with Low Enthalpy Water in the Rotorua Area of New Zealand," *Geothermics*, Special Issue 2, **2**(2), 1670–1675 (1970).

Coulbois, P., and Herault, J. P., "Conditions of Competitivity of Geothermal Domestic Heating," Second U.N. Symposium on the Development and Use of Geothermal Resources, San Francisco, May 20–29, 1975.

Delisle, G., et al., "Prospects for Geothermal Energy for Space Heating in Low Enthalpy Areas," Second U.N. Symposium on the Development and Use of Geothermal Resources, San Francisco, May 20–29, 1975.

Einarsson, Sveinn S., "Geothermal Space Heating and Cooling," Second U.N. Symposium on the Development and Use of Geothermal Resources, San Francisco, May 20–29, 1975.

Linton, A. M., "Innovative Geothermal Uses in Agriculture," Conference on Multi-purpose Use of Geothermal Energy, Klamath Falls, Oregon, October 7–9, 1974.

Lund, John, "Utilization of Geothermal Energy in Klamath Falls," Conference on Multi-purpose Use of Geothermal Energy, Klamath Falls, Oregon, October 7–9, 1974.

Luoviksson, Vilhjamur, "Multi-purpose Uses of Geothermal Energy," Second U.N. Symposium on the Development and Use of Geothermal Resources, San Francisco, May 20–29, 1975.

Ogle, William, "Alaskan Non Electric Geothermal Energy Needs and Possibilities," Conference on Multi-purpose Use of Geothermal Energy, Klamath Falls, Oregon, October 7–9, 1974.

Purvine, W. D., "Utilization of Thermal Energy at Oregon Institute of Technology, Klamath Falls, Oregon," Conference on Multi-purpose Use of Geothermal Energy, Klamath Falls, Oregon, October 7–9, 1974.

Reistad, G. M., "The Potential for Non-Electrical Applications of Geothermal Energy and Their Place in the National Economy," Second U.N. Symposium on the Development and Use of Geothermal Resources, San Francisco, May 20–29, 1975.

Storey, David M., "Geothermal Drilling in Klamath Falls, Oregon," Conference on Multi-purpose Use of Geothermal Energy, Klamath Falls, Oregon, October 7–9, 1974.

Thorsteinsson, Thorsteinn, "The Redevelopment of the Reykir Hydrothermal System in S.W. Iceland", Second United Nations Symposium on the Development and Use of Geothermal Resources, San Francisco, May 20–29, 1975.

Towse, Donald, "The Economics of Geothermal Heat as an Alternate Fuel," Lawrence Livermore Laboratory, Univ. California, Livermore, Calif. UCRL-77031, September 1975.

Wilson, R. D., "Use of Geothermal Energy at Tasman Pulp and Paper Company Limited—New Zealand," Conference on Multi-purpose Use of Geothermal Energy," Klamath Falls, Oregon, October 7–9, 1974.

Yuhara, K., and Sekioka, M. "Application of the Linear Programming to the Multi-purpose Utilization of Geothermal Resources," Second United Nations Symposium on the Development and Use of Geothermal Resources, San Francisco, May 20–29, 1975.

Zoega, Johannes, "The Reykjavik Municipal District Heating System," Conference on Multi-purpose Use of Geothermal Energy, Klamath Falls, Oregon, October 7–9, 1974.

Utilization, Mineral Recovery

Berthold, C. E., et al., "Process Technology for Recovering Geothermal Brine Minerals," Bureau of Mines Open File Report 35–75, February 4, 1975.

Lindal, Baldur, "Development of Industry Based on Geothermal Energy, Geothermal Brine and Sea Water in the Reykjanes Peninsula, Iceland," Second United Nations Symposium on the Development and Use of Geothermal Resources, San Francisco, May 20–29, 1975.

Lindal, Baldur, "Geothermal Energy for Process Use," Conference on Multi-purpose Use of Geothermal Energy, Klamath Falls, Oregon, October 7–9, 1974.

Werner, H. H., "Contribution to the Mineral Extraction from Supersaturated Geothermal Brines, Salton Sea Area, California," *Geothermics*, Special Issue 2, 2(2), 1651–1655 (1970).

CHAPTER 10

Comparison of Uses and Performance of Combined Systems

Currently, geothermal energy is used principally for the production of electricity and for space heating. The use is equally divided between the two as shown in Table 1.2. Along with space heating, geothermal energy is also used for the heating of greenhouses. The use of geothermal energy for process heat may be just beginning. It is used in Iceland for the drying of the product from a diatomaceous earth plant and in New Zealand for process steam at the Tasman pulp and paper mill. Greenhouse heating is a use that could be considered in one sense process heat and in the other space heat. Geothermal energy has also been proposed for the production of water, such as in the Imperial Valley region where a project is being undertaken by the United States Bureau of Reclamation. The recovery of minerals from geothermal brines has also been proposed and tried. The efforts to date have met with failure, particularly those in the Salton Sea region. The early efforts were limited, however, to the recovery of one or two particular minerals from the brine without any attempt to produce other materials or products, such as water or electricity. More recently, use of geothermal energy in a chemical complex for the production of minerals is being studied in Iceland. Because geothermal energy deposits are located in regions where mineral deposits will also be present, it is to be expected that geothermal energy will be used as process heat in conjunction with the process plants for mineral deposits. In addition, it would be expected that geothermal energy would be used for process heat for other process plants located in regions of geothermal energy deposits. It should be mentioned that geothermal brines might find particular application related to the recovery of minerals by *in situ* leaching processes where a hot fluid is required. The recovery of sulfur mentioned in Chapter 9 is an example of this.

All of the foregoing uses for geothermal energy, except the production of electricity, involve thermal heat as the product being supplied by geothermal energy. The production of water, for example, really involves the use of the thermal energy from the geothermal brines to operate some sort of multiple effect evaporator system for the production of fresh water. Thus the two principal uses for geothermal energy are the production of electricity and the production of thermal energy. The comparison of the net energy output as a function of brine temperature for the production of electricity and for the production of thermal energy shown in Figure 10.1 shows that the amount of thermal energy that can be produced is an order of magnitude greater than the amount of electricity that can be produced from a fixed quantity of geothermal brine. The electrical work output shown in this figure was obtained from Figure 8.27. In a cold climate, more heat can be extracted usefully than in a hot climate. The temperature range at which the thermal energy must be delivered for processing purposes depends on the use as described in Chapter 9. The thermal energy will be used with the greatest efficiency provided the entire temperature range from the delivery temperature of the wellhead brine to the surrounding temperatures can be used in a balanced fashion. There will be residential and commercial establishments wherever there is a processing plant; consequently, there will be requirements for space and water heating that occupy the lower temperature range as described in Chapter 9. Thus the energy of geothermal brines can be used from the wellhead temperature to a temperature near that of the environment.

As might be expected, the order of magnitude difference in useful energy output between thermal and electrical usage of geothermal energy as shown in Figure 10.1 significantly affects the resource's value. The meaning of this difference can be explored usefully in several different ways. Not too surprisingly, however, the conclusion is

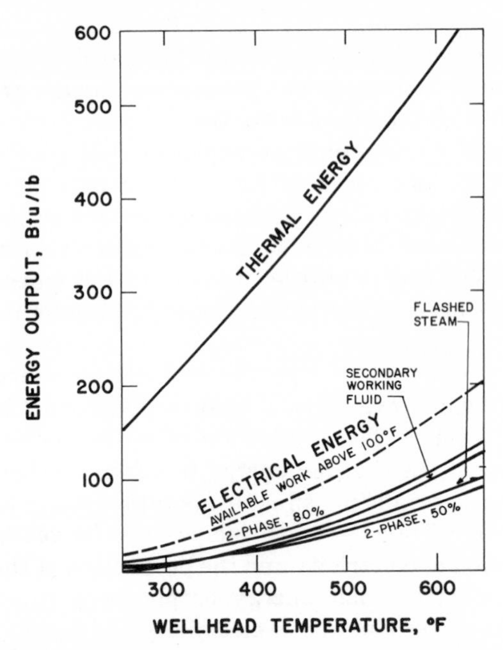

Figure 10.1 The productivity of geothermal brine for process and space heat compared with the production of electricity as a function of the wellhead temperature of the geothermal fluid. The theoretical limit of work, that is, available work, dashed line, and thermal energy utilization assume discharge to be 100°F at 1 atm. The other systems assume exhaust temperatures of 130°F.

similar, whether the analysis is done by maximizing the total resources for the benefit of the community over a long time period, by maximizing the profit of the reservoir owner, or by maximizing the profit of the industrial operations. Of course, this will not be true of all specialized interests because some will always gain at the expense of others.

To explore the significance of alternative uses of geothermal energy as well as to exemplify the analytical method for evaluating a particular case, several hypothetical examples will be discussed.

MAXIMUM BENEFITS FROM A FIXED RESOURCE

In this example, it is assumed that a fixed resource is available that is capable of delivering 100,000 lb/hr of oil and 10,000,000 lb/hr of geothermal brine. To provide the energy requirements for a community that is depending on this resource, two options are available. One, Case A, is to use the oil for producing electricity and to use the geothermal brine for providing space and process heat. Note that, in this example as well as in the others discussed later, only the electric, space, and process heat needs will be considered. The other option, Case B, is to use the geothermal fluid for generating electricity and the oil for providing space and process heat. To explore the complete range of geothermal fluids, each of these cases is compared in Table 10.1 for geothermal brines delivered at the wellhead with temperatures from 300 to 500°F.

For Case A, oil can be used for generating electricity in conventional power plants at an efficiency of 37%. Two-hundred-ten megawatt-hours of electricity are produced from the 100,000 lb of oil.

Table 10.1 Power Production in Megawatts from a Given Resource Supply Consisting of 10,000,000 lb/hr of Geothermal Brines and 100,000 lb/hr of Oil

Type of Power Production	Temperature Wellhead		
	300°F	400°F	500°F
CASE A. Using geothermal brine for producing space and process heat			
Electricity from oil[a]	210	210	210
Heat from geothermal[b]	586	879	1172
Total power	796	1089	1382
CASE B. Using geothermal brine for producing electricity			
Electricity from geothermal[c]	32	88	147
Heat from oil[d]	483	483	483
Total power	515	561	630
Ratio of Case A: Case B	1.5	1.9	2.2

[a]Oil to electricity at 37% thermal efficiency c.f. Table 8.1

[b]Rejection Temperature of 100°F

[c]From Figure 8.27

[d]Oil to process heat at 85% efficiency

Since geothermal brines delivered at the wellhead are providing thermal energy already in the desired form, no further processing is required except for correcting scale deposition problems as discussed earlier. Thus geothermal fluids will provide heat with an efficiency close to 100% for the thermal energy measured above the rejection temperature of the brine from the system. For the purposes of this comparison, it is assumed that this temperature is 100°F. In Iceland, for example, the rejection temperature is 105°F.

For Case B in which the geothermal fluid is used for producing electricity, the effect of the Carnot cycle efficiency on performance is quite evident. The efficiency with which the geothermal energy can be converted to electricity is taken from Figure 8.27. The work output per pound of brine is 11 Btu/lb at 300°F, 26 Btu/lb at 400°F, and 50 Btu/lb at 500°F. The thermal content of oil is converted with an efficiency of about 85% to process heat. This is a result of the combustion process that cannot transfer all of the heat to the heated fluid. In fact, if oil is used directly in homes for heating, the efficiency of conversion is much less, being closer to 65%.

Adding the total power produced for each of the cases shows that the use of geothermal energy for producing space and process heat provides not only the greatest total power, but also the greatest power of both electricity and heat individually. The ratio of the total power produced is 1.5 to 2.2 times greater using the geothermal fluid for heating and the oil for electricity than for the reverse case. Thus the use of geothermal fluid for space and process heat is the most efficient approach. The economics verify this as in the following case.

ECONOMICS OF UTILIZATION OF A FIXED RESOURCE

The market value of electricity and thermal energy delivered to residential or small commercial facilities is compared in Table 10.2. In this table, the value of the thermal energy has been corrected for the conversion efficiency of the oil or coal, as the case may be, to useful thermal energy. Thus the market value of delivered thermal energy from fossil fuel sources is about 50% higher than the cost of the heat of combustion of the raw material itself. In subsequent discussions, these values will be compared with geothermal energy, which delivers its thermal energy at close to 100% efficiency.

Consider as an example the case of the owner of a 10,000,000 lb/hr geothermal resource. The owner has the option of delivering the thermal energy for making electricity or process and space heat. The

Table 10.2 Delivered Price or Market Value of Electricity and Thermal Energy Expressed in Units of Both Million Btu and MWh.

	Values of Energy	
Type of Energy	$/$10^6$ Btu	$/MW-hr[a]
Electricity	11.7	40
Thermal Energy, 65% Efficiency	4.0	14

[a] $/MW-hr is equal to mils/kw-hr

relative profit that would be produced from the two cases will be calculated. As described in Chapter 8, geothermal brine at a temperature of 400 to 500°F can generate electricity that costs 12 to 18 mil/kWh delivered from the plant. On the other hand, fossil fuel-fired power plants produce electricity at a cost of about 25 mil/kWh. Thus the owner of the geothermal reservoir could expect to realize an additional profit of 7 to 13 mil/kWh compared with electricity produced from oil. That is, the use of the geothermal fluid for making electricity will yield a profit of 7 to 13 mils greater than the use of oil for making electricity.

On the other hand, if the geothermal fluid is used for providing space heat to an industrial community that is less than 10 mi distant, then the cost of delivering that thermal energy will be about 6 mil/kWh thermal. Since the delivered thermal energy will have a value of 14 mil/kWh as shown in Table 10.2, the profit from distributing the thermal energy will be about 8 mil/kWh.

A comparison of the profits that are obtained by the owner of the resource for the two different options, Case A and B of Table 10.3, shows that the profit potential from using a high temperature geothermal brine, that is, 500°F, for space and process heat is five times as great as for producing electricity. Electricity cannot be made profitably from a lower temperature fluid, that is, below 300°F. Thus the potential profit is substantially greater by selling the geothermal fluid for process and space heat rather than for generating electricity. This assumes, of course, that there is an adequate market within reasonable transport distance. As shown in Chapter 9, however, geothermal

Table 10.3 Profit Potential for Alternate Uses of Geothermal Fluid When Energy Is Sold at the Equivalent Price of a Fossil Fuel

	Wellhead Temperature		
	300°F	400°F	500°F
CASE A. Using 10,000,000 lbs/hr geothermal brine for producing space and process heat			
Thermal power, MW-hr	586	879	1172
Profit, $/MW-hr	8	8	8
Profit, $/hr	4688	7032	9376
CASE B. Using 10,000,000 lbs/hr geothermal brine for producing electricity			
Electrical power, MW-hr	-	88	147
Profit, $/MW-hr	0	7	13
Profit $/hr	0	616	1911
Ratio of profits Case A:B	∞	11	5
CASE C. Using 1,000,000 lbs/hr geothermal steam for producing space and process heat			
Thermal power, MW-hr	317	323	323
Profit, $/MW-hr	8	8	8
Profit, $/hr	2536	2584	2584
CASE D. Using 1,000,000 lbs/hr geothermal steam for producing electricity			
Electrical power, MW-hr	-	62	76
Profit, $/MW-hr	0	7	13
Profit, $/hr	0	434	912
Ratio of profits Case C:D	∞	6	3

fluid can be economically transported a considerable distance, depending on how much power is being transmitted. Iceland has shown that transport distances of 10 mi can be profitable. Chapter 9 shows that in the United States it is potentially feasible to transport fluids for hundreds of miles depending on the power level.

In this comparison, the profit associated with the distribution of electricity has not been included because it is a difficult number to establish properly. If this were done and the total profit calculated, the conclusions would be similar, however.

Ten million pounds per hour of geothermal brine represents the output of 20 to 30 geothermal wells and is about the optimum utilization for a given central collecting plant from a number of wells. This applies whether the collection point is a pumping station for distribution of thermal energy or whether it is a collection point for the production of electrical power. A steam-producing geothermal resource produces about 1,000,000 lb/hr of steam from the same number of wells. An economic comparison for the utilization of geothermal steam similar to that made for geothermal brine is shown by Cases C and D in Table 10.3. The profit from thermal energy is three to six times that for electrical production. Thus for the case of a steam reservoir, the conclusions regarding the economics of utilization are the same as for brine.

The foregoing discussion assumes that the transport distance for the thermal energy is less than 10 mi. If there is no market for the thermal energy within reasonable distance, then of course these conclusions will be altered. Also in an actual case, there will be many other specific factors that must be analyzed and might alter the conclusions.

COMBINED THERMAL AND ELECTRICAL PRODUCTION

A geothermal brine processing plant that is converting brine thermal energy into electricity will always discharge heat. Because of the low temperature of geothermal brines, the cycle efficiency is low and the discharged heat is the major portion of the energy flowing out of the plant. Consequently, any use of this thermal energy will greatly increase the efficiency of geothermal energy utilization. This means, however, that the heat discharged from the plant must be at a higher temperature than would be desired for obtaining maximum electrical power.

A simplified schematic representation of a plant processing geothermal brine to produce electrical work and discharge thermal energy is

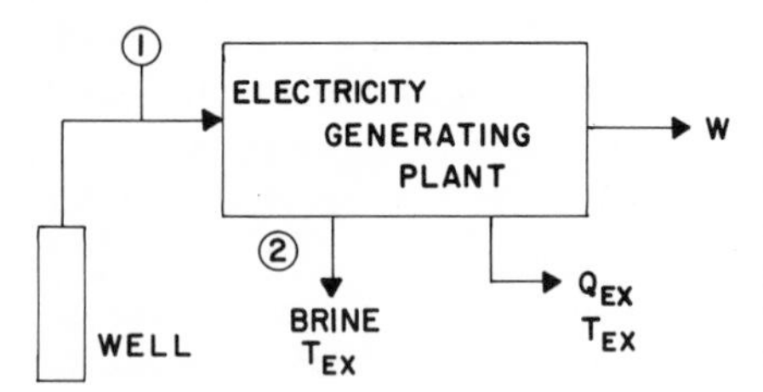

Figure 10.2 Simplified schematic of the system for converting the thermal energy of geothermal brine in an electricity-generating plant to work and showing the thermal energy discharged at the rejection temperature of the system.

shown in Figure 10.2. Applying the first law of thermodynamics, Equation 6.4, to the process plant shown in Figure 10.2 results in

$$Q_{EX} = (h_1 - h_2) - W_A \tag{10.1}$$

As shown in Chapter 8, the work produced from geothermal brine, whatever process system is used, is given by

$$W_A = e[h_1 - h_2 - T_{EX}(s_1 - s_2)] \tag{10.2}$$

The electrical work output is obtained by multiplying the mechanical work W_A by the generator efficiency

$$W_E = e_G W_A \tag{10.3}$$

Inserting W_A from Equation 10.2 into Equation 10.1 and dividing the exhaust thermal energy by the work output gives the ratio of thermal energy that can be utilized to the actual work output:

$$\frac{Q_{EX}}{W_E} = \frac{[(h_1 - h_2) - e(h_1 - h_2 - T_{EX}\{s_1 - s_2\})]}{e_G e[h_1 - h_2 - T_{EX}(s_1 - s_2)]} \tag{10.4}$$

This equation shows that the amount of rejected heat is a strong function of the exhaust temperature of the system because the net work output per pound of brine is relatively small. For maximum electrical power output, the exhaust temperature should be the lowest consistent with the surrounding environment. This will generally be somewhere between 70°F for cold climates and 130°F for warmer climates. If the rejected thermal energy is to be utilized for space or process heat, it must be rejected at a higher temperature to be useful. A temperature of 180°F is probably adequate for most space and water heating purposes. Slightly higher temperatures would be necessary for processing purposes. In Iceland, for example, where the geothermal brine is used solely for heating purposes, it is mixed with cool water to a temperature of 180°F before pumping through the distribution system.

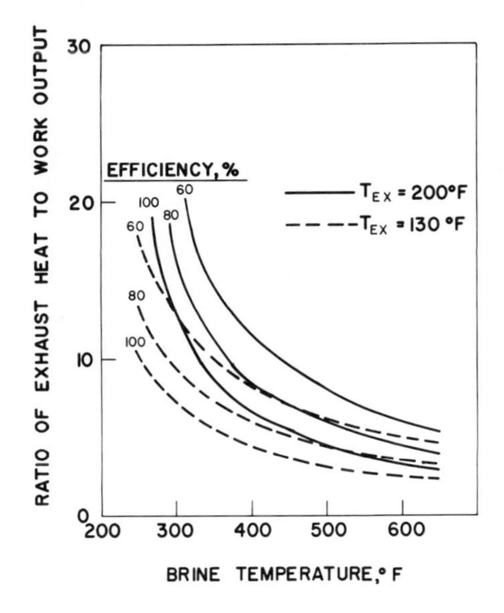

Figure 10.3 The ratio of the exhaust heat to the electrical production of a geothermal electricity generating plant as a function of the brine temperature for various exhaust temperatures from the generating plant. The solid line is for an exhaust temperature of 200°F and the dashed line is for an exhaust temperature of 130°F. The reject temperature of the brine is assumed equal to the exhaust temperature of the system. Consequently, the brine will contain additional useful thermal energy above that shown. The above curves are based on various turbine or system efficiencies as shown on each of the appropriate curves. The electrical generator efficiency is assumed to be 95%.

The ratio of the exhaust heat that is available for thermal purposes to the electrical work output as given by Equation 10.4 is shown in Figure 10.3 for exhaust temperatures of 130 and 200°F. This figure shows that the electrical work output will amount to 5 to 10% of the thermal energy output for brine temperatures above 400°F but falls dramatically for temperatures below that. The district heating system in Reykjavik, Iceland, requires a pumping power that amounts to 2% of the total thermal power delivered by the system. Thus the electrical power output that can be obtained from geothermal brine in a combined system of power and thermal production will be sufficiently large that there will be some power produced in excess of that required for pumping purposes.

The Nesjavellir Combined System

The previous considerations are exemplified by a comparison of various methods for utilizing the Nesjavellir geothermal field in Iceland. This reservoir can be utilized for producing electrical power only, for thermal power only, or for a combined thermal and electrical power production system (Zoega, 1974) as summarized in Table 10.4. The maximum electrical power will be obtained from the geothermal energy by using a condenser operating at 90°F. The electrical power production in this case is 69 MW from a well production containing

Table 10.4 Comparison of Methods for Utilizing Nesjavellir Geothermal Field

	Condenser Temperature[a] (°F)	Electrical Power[a] (MW)	Heat Power[a] (MW)	Total Power[a] (MW)	Relative Value[b] ($/hr)
Well Production[c]	--	--	--	696	--
Heating only[d]	--	--	531	531	7434
Condensate for heat	180	42	505	547	8750
Low temp. condenser	90	69	289	358	6806

[a]Source. Zoega, 1974.

[b]Fossil Fuels deliver electricity at 37% thermal efficiency and process heat at 85% thermal efficiency, so that electrical power is 3 to 4 times as valuable as heat on a basis of minimizing follis fuel consumption.

[c]Brine delivered at 500°F and discharged at 105°F.

[d]Heat delivered above 105°F by heating cold water from 40°F to 226°F.

696 MW of thermal power measured as the thermal energy content of the brine above 90°F. Some of the brine is diverted for providing thermal power in the amount of 289 MW, and the rest of the brine is utilized for producing electricity at 69 MW. The resulting total power is 358 MW of energy.

For utilization of the geothermal reservoir for space heating, the hot brine is diluted with cold water to produce a thermal fluid for distribution at 180°F. This dilution with colder water is necessary to prevent precipitation of salts in the thermal distribution system as mentioned in Chapter 5 under "Techniques for Controlling Deposition of Scale from Geothermal Brines." Thus the use of the geothermal reservoir for heating purposes actually results in the delivery of only 531 MW of thermal power. The loss between the 696 MW available and the 531 MW delivered is due to the heating of the cold 40°F water for dilution to the return temperature of 105°F from the district heating system.

To utilize the rejected heat from the electrical production facility for space heating, the condenser temperature is increased from 90 to

180°F. This results in a reduction of the electrical power from 69 to 42 MW. The rejected heat at 180°F is available to produce thermal power of 505 MW compared to 289 MW for the 90°F condenser case. Thus for a small reduction in electrical power production, a substantial amount of thermal energy for space heating is provided. The total power delivered in this case then amounts to the sum of 42 MW of electrical power and 505 MW of thermal power for a total of 547 MW of power. As shown in Table 10.4, the maximum delivered power is obtained by using a combined electrical and thermal power production facility in which the rejected heat from the electrical plant is utilized at a condenser temperature of 180°F for thermal purposes. Because of the three or four to one relative value of electricity to thermal energy, the benefit obtained by a combined system relative to a thermal system only is greater than appears from an evaluation of the relative performance on the basis of total power only. This is shown in Table 10.4 where the relative value of the products is computed using the values of thermal power and electrical power given in Table 10.2. This comparison shows that the economic benefit from this field is maximized by using a combined thermal and electrical production plant. It also shows, however, that because of the higher value of the electricity, the production of electrical power alone in a separate facility from thermal production is not as disadvantageous as indicated by the total power level.

Use of Two-Phase Expansion Machines for Topping Cycles

For the purpose of using the upper temperature range of geothermal brines for producing electricity, called a topping cycle, it would be advantageous to have available inexpensive capital cost process equipment for accomplishing the energy transformation. The thermal energy rejected in the condenser of a topping cycle must be at a temperature of about 200°F or more to be useful. Consequently, the temperature and therefore pressure range of the topping cycle will be limited. Because the two-phase expansion machines described in Chapter 7 operate well over a low expansion ratio, they may be particularly suited to topping cycles. In addition, they are capable of operating with high scale-depositing brines such as the Salton Sea brines that cannot now be efficiently processed except by thermodynamically inefficient processes of flashing and washing of the flash steam.

Use of a two-phase expansion machine in a topping cycle yields increased profits of 10 to 70% depending on the transmission distance required for the thermal energy distribution. A significant effect of adding a two-phase topping process is to increase the transport distance to the industrial site that would be economically attractive. Therefore development of two-phase expansion machines may contribute to the successful implementation of combined electrical production and thermal production.

COMBINED GEOTHERMAL RESOURCE AND MINERAL RESOURCE USING GEOTHERMAL ENERGY FOR PROCESS HEAT

To understand the value of geothermal energy for process heat, consider the hypothetical case of a process plant producing 100 ton/hr of product located 10 mi from a geothermal reservoir. The process might be the conversion of a mineral resource into a marketable commodity and is assumed to require 5,000,000 Btu of thermal energy above 200°F for every ton of product produced. If fuel oil is available for providing the thermal energy at a cost of $3.00/1,000,000 Btu, the cost of the thermal energy is $15.00/ton of product. In this example, two cases will be considered. Case A is an operation with a gross profit of $5.00/ton of product using conventional fuel. Case B assumes that the plant loses $5.00/ton of product using conventional fuel.

For each case, two options for utilizing the geothermal energy are considered. One is to process the material from the plant using conventional fuel and use the geothermal energy for generating and marketing electricity. The other is to use the geothermal energy for processing the product from the plant. The question then is which of the two options produces the maximum profit from the given resource.

Assuming that ownership of the reservoir and the processing plant, as well as perhaps of oil reservoirs at some other location, are in the hands of one owner makes the analysis somewhat simpler. The geothermal reservoir is assumed to deliver brine at the wellhead at a temperature of 300°F and a flow rate of 420,000 lb/hr for each well and to be located 10 mi from the processing plant. Since only the thermal energy above 200°F can be used by the process, the available energy per pound of brine is 100 Btu. This means that each well will produce 42,000,000 Btu/hr of usable thermal energy. Since 500,000,000 Btu/hr are required to produce 100 ton/hr of product, 12 wells will be required.

The cost of transporting the geothermal brine 10 mi, using Equation 9.19 and an annual cost of 20% of the total capital, is $3.13/1,000,000 Btu transported. If the average well depth is 4000 ft, the cost per 1,000,000 lb of brine produced is $20.00 from Table 9.8. Since the brine contains 100 Btu/lb and since 5,000,000 Btu are required per ton of product, the cost of the well and collection system per ton of product is $1.00/ton. Combining the cost of the transmission system and the cost of the well and collection system the total cost is $4.13/ton of product for the geothermal energy delivered to the plant. Since the cost of fuel oil is $15/ton of product, the savings amount to $10.87/ton of product. Thus as is shown in Table 10.5, the profit for Case A using geothermal energy as process heat is the $5.00 original profit using fuel oil plus the additional $10.87 savings resulting in a total profit of $15.87/ton of product. For Case B, profit is computed as shown in Table 10.5 and amounts to $5.87/ton of product. Since 100 ton/hr of product are produced, the respective profit for Cases A and B are $1587.00 and 5.87/hr.

Table 10.5 Profit Utilizing a Geothermal Resource for a Process to Produce 100 ton/hr of Product Compared with the Profit from Use of the Geothermal Resource for Electricity and Using Fuel Oil for Process Heat

Geothermal Resource Specifications		
Location	10 miles from process plant	
Well depth	4000 feet	
Well head temperature	300°F	
Well flow rate	420,000 lbs/hour	
Process Specifications		
Plant size	100 tons/hour product	
Thermal requirements above 200°F	5×10^6 Btu/ton product	
Cost of fuel oil per ton of product	$15/ton product	
Economic element	Case A	Case B
Product profit using fuel oil, per ton	$ 5	$ (5) loss
Option 1 Geothermal Energy Used for Process Heat		
Cost of fuel oil minus cost of geothermal, per ton	$ 10.87	$ 10.87
Gross profit per ton product	$ 15.87	$ 5.87
Gross profit per hour, process plant	$1587	$587
Option 2 Geothermal Energy Used for Producing Electricity		
Profit from electricity production	$ 15	$ 15
Profit from product	500	-
Gross profit per hour, process plus electricity	$515	$ 15

The profit that is made by using geothermal energy for electricity may be estimated by assuming that the profit is the difference between the cost of electricity produced from fuel oil and that produced by geothermal energy. This amounts to about 7 to 13 mil/kWh at 400°F. This does not include the profit from distributing electricity that would be retained by the distributing and marketing companies, that is, the electric utilities. Since the geothermal brine was assumed to be available at 300°F, the thermal energy of the brine would be converted to electricity with an efficiency factor of about 5%. Since the total assumed output is 500,000,000 Btu/hr, the amount of electricity that could be produced from this thermal energy content is about 5 MWh. At a profit of 3 mil/kWh for 300°F brine, this means that the profit from the sale of geothermal energy for the purpose of making electricity would at best be $15.00/hr. For this option, shown in Table 10.5, the total profit amounts to $515.00/hr. This is one-third the profit of $1587.00 when the geothermal energy is used for process heat.

This example shows the importance of using geothermal energy for process heat purposes. It also exemplifies the increase in profits that will be experienced by an owner of energy resources and processing plants who uses the geothermal energy for process heat rather than selling it for electrical production.

SUMMARY

The production of both electricity and thermal energy from a geothermal resource maximizes the total value of the products. This may be accomplished by making electricity from a topping cycle and using the heat from the condensing section of the electrical plant to provide space and process heat. For processes that require thermal energy at a higher temperature level, a combined electrical production and process heat facility may not be possible.

A geothermal resource is more valuable and may return a greater profit when used for space and process heat because of the low Carnot cycle efficiencies resulting from the low temperature of the brine. The greater value as a heating source also results from the lower brine temperature giving nearly reversible heat transfer to the process or heating system. On the other hand, fuel oil or conventional fuels provide heat at a much higher temperature; therefore, when used directly for space or process heat purposes without prior conversion to electricity, they have a high irreversible loss as a result of the large temperature difference. As a result of these two factors, the use of geothermal energy for space and process heat returns a greater profit than its use for making electricity.

NOMENCLATURE

e efficiency
h specific enthalpy, Btu/lb
Q_{EX} heat exhausted from the process per pound of brine processed, Btu/lb; positive for outflowing heat
s specific entropy, Btu/lb
W specific work per pound of brine processed, Btu/lb

Subscripts

A actual mechanical work produced
E electrical
EX exhaust

REFERENCE

Zoega, Johannes, "The Reykjavik Municipal District Heating System," Conference on Multi-purpose Use of Geothermal Energy," Klamath Falls, Oregon, October 7–9, 1974.

Index